Improving Integration and Synchronization of Space Acquisition and Fielding

BONNIE L. TRIEZENBERG, WILLIAM SHELTON, MEGAN MCKERNAN, SARAH W. DENTON, JAMES DIMAROGONAS, BRIAN DOLAN, SHANE MANUEL, GWEN MAZZOTTA, SYDNE J. NEWBERRY, LAURINDA L. ROHN, KAREN SCHWINDT, YULIYA SHOKH, JORDAN WILLCOX

Prepared for the Department of the Air Force
Approved for public release; distribution is unlimited.

RAND PROJECT AIR FORCE

For more information on this publication, visit **www.rand.org/t/RRA1735-1**.

About RAND

The RAND Corporation is a research organization that develops solutions to public policy challenges to help make communities throughout the world safer and more secure, healthier and more prosperous. RAND is nonprofit, nonpartisan, and committed to the public interest. To learn more about RAND, visit www.rand.org.

Research Integrity

Our mission to help improve policy and decisionmaking through research and analysis is enabled through our core values of quality and objectivity and our unwavering commitment to the highest level of integrity and ethical behavior. To help ensure our research and analysis are rigorous, objective, and nonpartisan, we subject our research publications to a robust and exacting quality-assurance process; avoid both the appearance and reality of financial and other conflicts of interest through staff training, project screening, and a policy of mandatory disclosure; and pursue transparency in our research engagements through our commitment to the open publication of our research findings and recommendations, disclosure of the source of funding of published research, and policies to ensure intellectual independence. For more information, visit www.rand.org/about/research-integrity.

RAND's publications do not necessarily reflect the opinions of its research clients and sponsors.

Published by the RAND Corporation, Santa Monica, Calif.

Library of Congress Cataloging-in-Publication Data is available for this publication.

ISBN: 978-1-9774-1209-6

Cover credits, clockwise from top right: US Army; U.S. Army; Composite image: Satellite: PhonlamaiPhoto/Getty Images/iStockphoto, Drone: AlexLMX/Getty Images/iStockphoto, Sky: huafires/Getty Images/iStockphoto.

About This Report

One of the key goals of the United States Space Force (USSF) is to deliver space capabilities on a faster timeline to outpace adversary threats. However, delivering an end-to-end space capability requires the integration and synchronization of multiple elements (e.g., space systems, ground command and control systems, and user equipment) whose development and fielding are managed by not just the Department of the Air Force but also by other military departments and federal agencies. Additionally, to increase the resilience and warfighting capacity of the space enterprise, USSF may need to work across the Department of Defense and other government agencies to integrate and synchronize complementary capabilities across multiple mission areas (e.g., overhead persistent infrared space systems to support missile warning and missile defense). In its new role, the Office of the Assistant Secretary of the Air Force for Space Acquisition and Integration (SAF/SQ) has made improving the integration and synchronization of the delivery of space capabilities a priority. The objective of this report is to recommend ways that USSF can facilitate the integration and synchronization of space acquisition and fielding to support the timely delivery of end-to-end space capabilities to the joint force.

The research reported here was commissioned by USSF, SAF/SQ, and conducted within the Resource Management Program of RAND Project AIR FORCE as part of a fiscal year 2022 project, "Improving Integration and Synchronization of Space Acquisition and Fielding."

RAND Project AIR FORCE

RAND Project AIR FORCE (PAF), a division of the RAND Corporation, is the Department of the Air Force's (DAF's) federally funded research and development center for studies and analyses, supporting both the United States Air Force and the United States Space Force. PAF provides the DAF with independent analyses of policy alternatives affecting the development, employment, combat readiness, and support of current and future air, space, and cyber forces. Research is conducted in four programs: Strategy and Doctrine; Force Modernization and Employment; Resource Management; and Workforce, Development, and Health. The research reported here was prepared under contract FA7014-22-D-0001.

Additional information about PAF is available on our website:
www.rand.org/paf/

This report documents work originally shared with the DAF on August 3, 2023. The draft report, dated September 2022, was reviewed by formal peer reviewers and DAF subject-matter experts.

Acknowledgments

The RAND PAF research team thanks the many subject-matter experts from USSF, the United States Space Command, and the U.S. Air Force, along with others within the Department of Defense, intelligence community, Department of Commerce, federally funded research and development centers, and industry who provided valuable insights for this research. All participants spoke on a nonattribution basis and although we cannot mention them by name, we are grateful for their inputs.

A special thanks goes to the sponsor of this research, Maj Gen Steven P. Whitney, Military Deputy, SAF/SQXP. We also thank Erin Carper, chief, Acquisition Policies and Processes Division, SAF/SQXP, for her insights and assistance as our action officer. Lt Col David Corder, SAF/SQXP, provided us additional perspective and assistance throughout the research. Finally, we thank Brian Shannon, Col Franklin D. Gaillard II, and Chris Beauregard for their feedback throughout the project and for helping us connect with key subject-matter experts.

We thank Cynthia Cook and Jonathan Wong, our formal peer reviewers, who helped improve the quality of this document through their comments and suggestions. Yool Kim also provided useful inputs during the research process that helped inform the team. Christopher Adams was invaluable in helping us understand the network analysis visualization software. We thank Tandrea Reeves, Evan Smith, and Maria Falvo for their administrative assistance during this effort. Last, we thank Stephanie Young (director), Anna Jean Wirth (associate director), and Patrick Mills (former associate director) of the Resource Management Program for their leadership and useful feedback.

Summary

Issue

As part of a new role, the Office of the Assistant Secretary of the Air Force for Space Acquisition and Integration (SAF/SQ) has identified improving vertical (e.g., with space, ground, and user segments) and horizontal (e.g., with the stakeholder community—including the other armed services—other government agencies, and industry) integration and synchronization as a priority. To assist with this priority, the RAND Corporation's Project AIR FORCE (PAF) was tasked with defining challenges and recommended ways the United States Space Force (USSF) can facilitate the integration and synchronization of space acquisition and fielding to support the timely delivery of end-to-end space capabilities to the joint force.

Approach

We undertook a broad literature review of existing research on U.S. government space-affiliated organizations to ensure that the analysis leveraged and built on prior insights. To supplement the literature review and to determine the current state of space acquisition, we also conducted more than 60 semistructured interviews with subject-matter experts and senior decisionmakers throughout the Department of Defense, other federal government agencies, and the private sector.[1] Finally, to understand the current state of the USSF acquisition ecosystem (i.e., the connections between the various stakeholders in space acquisition), we mapped approximately 90 different stakeholders in USSF acquisition and the relationships among those organizations.

Key Findings

Key findings from our research are:

- Budget instability at the program level and overall budget inflexibility make it difficult to execute and deliver integrated and synchronized capabilities.
- USSF's ability to orchestrate and deliver a resilient and interoperable space architecture is unclear.
- Internal USSF organizations do not appear to be aligned and communication channels are adversely affected.
- Significant experimentation is ongoing within multiple organizations devoted to innovation, but this experimentation is not visible centrally or across organizations.

[1] This report was completed in August 2023, and the term current in this report should be understood as "as of 2022."

- USSF needs to harness and transition the outcomes of experiments and innovation into programs of record or into operations.

Recommendations

Budget Resiliency

To improve budget resiliency, we recommend that USSF take the following actions:

- Create resource-loaded roadmaps of mission capability architectures and conduct sensitivity analyses to better understand and prepare for the impacts of budget actions, such as continuing resolutions, changes to funding levels, or program cancellations.
- Implement a cultural change that values more-open communication with Congress and builds alignment across USSF (field commands and acquisition community), Department of the Air Force, Pentagon, and the greater U.S. space enterprise.

Architectural Focus

To improve architectural focus, we recommend USSF take the following actions:

- Define what constitutes architecture and use this definition to document USSF mission statements and charters.
- Educate the USSF workforce on architecture vision, definitions, responsible offices, and relationships among systems and missions within the architecture.
- Clarify the Space Warfighting Analysis Center's and Space Systems Integration Office's roles in defining system of systems (SoS) architecture and in advising the Service Acquisition Executive and SAF/SQ. While SAF/SQ may choose to outsource its needs for SoS engineering talent, advice should not be filtered through an external organization.
- Conduct a workforce study to determine how best to create and sustain an organic SoS and systems engineering expertise within USSF.

Alignment of United States Space Force Organizations

To better align USSF organizations, we recommend USSF take the following actions:

- Formalize and approve mission statements and charters to clarify and document organizational roles in space acquisition.
- Harmonize the functional constructs used in describing space operations and acquisitions to the extent reasonable and standardize mission area terminology.
- Continue to improve operator and acquirer communication processes.

Harnessing Space Innovation and Experimentation

To better harness ongoing innovation and experimentation, we recommend USSF take the following actions:

- Require organizations involved in space acquisition to improve visibility into their efforts and ensure that capability roadmaps include all experiments and innovations.

- Implement a cultural change that values enterprise and mission success over personal or individual program success.

Transitioning Technology from Acquisition to Operations

To more effectively transition acquired items into operations, we recommend USSF take the following actions:

- Create a clear vision between acquisition and user communities as to how and when technologies should be integrated into the broader capability architecture.
- Ensure there is funding for transitioning technology to operations.
- Develop and formalize a process to incorporate warfighter feedback into acquisition decisionmaking.
- Train and familiarize acquisition professionals on tailoring the different acquisition pathways, depending on type of capability, mission, and so forth.

Contents

Figures and Tables

Figures

Tables

Chapter 1. Introduction

Background

As the newest service of the Armed Forces, one of the key goals of the United States Space Force (USSF) is to deliver space capabilities on timelines that outpace adversary threats.[1] However, delivering an end-to-end space capability requires the integration and synchronization of multiple elements (e.g., space systems, ground command and control [C2] systems, and user equipment) whose development and fielding are managed by not just the Department of the Air Force (DAF) but also by other military departments and federal agencies. Additionally, to increase the resilience and warfighting capacity of the space enterprise, USSF may need to work across the Department of Defense (DoD) and government agencies to integrate and synchronize complementary capabilities in multiple mission areas (e.g., overhead persistent infrared space systems to support missile warning and missile defense). The Government Accountability Office (GAO) has noted the importance of synchronization of the space enterprise:

> . . . DoD is at a crossroads with space, as it is seeking to bolster its protection of space systems. . . . It is important that DoD address barriers that stand in the way of making these transitions, such as leadership fragmentation and lack of synchronization between satellites and their corresponding ground acquisition programs.[2]

From our research, we came to understand the issue of integration and synchronization as not simply applying to the acquisition of technical solutions (e.g., space systems integrated with ground systems) but more broadly as the integration and synchronization of human capital, partnerships, data, and technical solutions. In this, we reflect the philosophy of the Joint All-Domain Command and Control (JADC2), which will integrate and synchronize C2 in multiple domains through five lines of effort (LOEs) to field a fully operational capability

> to organize and guide actions to deliver materiel and non-materiel JADC2 capabilities. The LOEs are: (1) establish the JADC2 Data Enterprise; (2) establish the JADC2 Human Enterprise; (3) establish the JADC2 Technical Enterprise; (4) integrate Nuclear C2 [command and control] and

---

[1] Public Law 116-92, National Defense Authorization Act for Fiscal Year 2020, December 20, 2019, specified the formation of a space force separate from the Air Force. USSF was stood up on December 20, 2019, with the signing of the United States Space Force Act as part of the fiscal year 2020 National Defense Authorization Act (NDAA).

[2] Jon Ludwigson, *Space Acquisitions: Challenges Facing DoD as It Changes Approaches to Space Acquisitions*, testimony before the Subcommittee on Strategic Forces, Committee on Armed Services, House of Representatives, Government Accountability Office, GAO-16-471T, March 9, 2016, p. 1.

> Communications (NC2/NC3) with JADC2; and (5) modernize mission partner information sharing.[3]

The Office of the Assistant Secretary of the Air Force for Space Acquisition and Integration (SAF/SQ) has defined improving the integration and synchronization of the delivery of space capabilities as a priority. With the confirmation and swearing in of Frank Calvelli in May 2022, the USSF also has a Service Acquisition Executive (SAE), who is responsible for managing and controlling USSF acquisition to ensure successful and efficient operations. The SAE is subject to the direction of the Under Secretary of Defense for Acquisition and Sustainment (USD/A&S) on acquisition matters and the authority, direction, and control of the Secretary of the Air Force.[4] Therefore, acquiring and delivering end-to-end space capabilities in a timely fashion to meet mission requirements is an SAE responsibility.

Integration and Synchronization Challenges

Integration and synchronization challenges have existed for decades within the DoD space acquisition community. However, as RAND Corporation researchers concluded in a 2021 report,

> more than other warfighting domains, delivering warfighter services from space necessitates highly orchestrated horizontal and vertical synchronization. Horizontal synchronization is necessary to enable interaction among legacy and newer systems, cross-service participation in space efforts, and critical involvement of other agencies. Vertical synchronization is critical for ground and space systems interoperability, an area that has suffered inadequate attention in the past.[5]

GAO has reported frequently on the vertical synchronization problem, which includes the synchronization of a satellite, ground system, and user terminals, and has resulted in "ground system delays . . . so lengthy that satellites sometimes spend years in orbit before key capabilities can be fully used."[6] GAO discussed space C2 vertical synchronization challenges:

> the Space C2 Enterprise Manager has management responsibility—but not authority—over multiple development efforts included in the Space C2

---

[3] DoD, *Summary of the Joint All-Domain Command and Control (JADC2) Strategy*, March 2022b, p. 2.

[4] The USSF SAE position was established in the fiscal year 2020 NDAA with a requirement to be filled by October 1, 2022. Congress created this position in response to concerns that there were too many organizations overseeing space procurements. The SAE was confirmed in May 2022, late in the period during which we conducted our research. See Sandra Erwin, "New Head of Space Force Acquisitions Looks to Get Back to Basics," *SpaceNews*, June 24, 2022e; U.S. Code, Title 10, Section 3103, Civilian Management of the Defense Acquisition System.

[5] William Shelton, Cynthia R. Cook, Charles Barton, Frank Camm, Kelly Elizabeth Eusebi, Diana Gehlhaus, Moon Kim, Yool Kim, Megan McKernan, Sydne Newberry, and Colby P. Steiner, *A Clean Sheet Approach to Space Acquisition in Light of the New Space Force*, RAND Corporation, RR-A541-1, 2021.

[6] Cristina T. Chaplain, *Space Acquisitions: DoD Continues to Face Challenges of Delayed Delivery of Critical Space Capabilities and Fragmented Leadership*, testimony before the Subcommittee on Strategic Forces, Committee on Armed Services, U.S. Senate, Government Accountability Office, GAO-17-619T, May 17, 2017, p. 1.

> enterprise. For example, technology maturation and risk reduction activities are divided across three program offices, managed by two program executive officers, and reliant upon multiple sources of information. This division of work is being done in part because the various organizations have areas of expertise that the program was hoping to leverage. However, such distribution of activities among many organizations can result in synchronization and coordination challenges.[7]

Likewise, horizontal synchronization challenges have also been difficult to overcome. This includes the synchronization of the various stakeholders in the space community who are either acquiring, building, or using space capabilities, such as DoD, the intelligence community (IC), other federal government agencies, the commercial sector, and allies and partners. In 2016, GAO found that DoD space leadership was fragmented, which has contributed to synchronization and integration challenges:

> We identified approximately 60 stakeholder organizations involved in space acquisitions. There are eight organizations with space acquisition management responsibilities. While the Air Force has responsibility for most military space acquisitions, the other military services have their own space efforts as well. The NRO [National Reconnaissance Office] collaborates with DoD as it develops space systems for DoD and the IC. Oversight is spread across 11 offices within the Air Force, Office of the Secretary of Defense (OSD), the IC, and OMB [Office of Management and Budget]. . . . Lastly, six DoD organizations from the services, U.S. Strategic Command, and the Office of the Joint Chiefs of Staff are involved in setting requirements for defense space programs.[8]

RAND research on space acquisition found that the seam between the acquisition and operational communities is artificial and that removing it would improve integration and synchronization.[9] This finding was reinforced in our discussions with space subject-matter experts (SMEs) from both communities during this project. Both communities must have a deep understanding of the relevant technology, its advancement, and its application, and must have no physical requirements that separate performance of their respective core functions. If these communities operated from a shared background or job code, they would be able to communicate more effectively. Greater specialization—if needed—could be achieved later in an individual's career, but the shared bond would remain.

Recent Changes That Affect the Synchronization Challenge

Since 2016, the landscape we just described has changed with the reinitiation of U.S. Space Command (USSPACECOM) as a geographic combatant command, the establishment of the

---

[7] Chaplain, 2017, p. 1.

[8] Cristina T. Chaplain, *Defense Space Acquisitions: Too Early to Determine If Recent Changes Will Resolve Persistent Fragmentation in Management and Oversight*, Government Accountability Office, GAO-16-592R, July 27, 2016, p. 15.

[9] Shelton et al., 2021.

Space Development Agency (SDA) and the Space Rapid Capabilities Office (Space RCO), and the ongoing maturation of USSF as a new service. Although these changes have promise, there are still many stakeholders in space and new lines of communication and interaction are still forming. As we will discuss later in this report, the timing of our research provides a snapshot of those interactions, and several of our recommendations center on actions USSF should take *now* to best align itself for a more synchronized future.

Simultaneous with the ongoing reorganization of space acquisition organizations, DoD implemented the Adaptive Acquisition Framework (AAF) to improve acquisition timelines. This revision to DoD Directive (DoDD) 5000.01 created a set of tailored acquisition pathways.[10] While USSF uses the major capability acquisition pathway for some traditional acquisition programs, it uses the middle tier of acquisition (MTA) pathway (e.g., rapid prototyping and fielding) and software acquisition pathway (e.g., agile software practices) to shorten the timelines from requirements definition to fielding capabilities. The AAF is just one of many tools that can be leveraged to improve the acquisition of space systems (e.g., whether the fielded space systems deliver a timely, integrated, and synchronized warfighter capability for the joint fight).

Complexity of What the Space Community Acquires

The complexity of space acquisitions is varied and defined along multiple dimensions. At one end of the complexity spectrum is commercially available or otherwise ready-to-use technology. See Figure 1.1 for an illustration of the range of complexity.

---

[10] DoDD 5000.01, *The Defense Acquisition System*, Office of the Under Secretary of Defense for Acquisition and Sustainment, September 9, 2020, incorporating Change 1, July 28, 2022; DoD Instruction (DoDI) 5000.02, *Operation of the Adaptive Acquisition Framework*, Office of the Under Secretary of Defense for Acquisition and Sustainment, January 23, 2020, incorporating Change 1, June 8, 2022.

Figure 1.1. Varied Complexity in Space Acquisitions

Complexity: TRL, Integration, Assuredness, Production Scheme

- Commercially available or otherwise ready technology
- Single space-centered user community
- Little integration with other systems
- Relaxed mission assuredness requirements
- Large production runs of standardized and nonsensitive systems

Weather as a service

- Advanced technology requiring significant novel development
- Broad user community across domains and organizations
- Extensive and critical integration challenges
- Stringent mission assuredness requirements
- Small production runs of highly complex, atypical, and sensitive systems

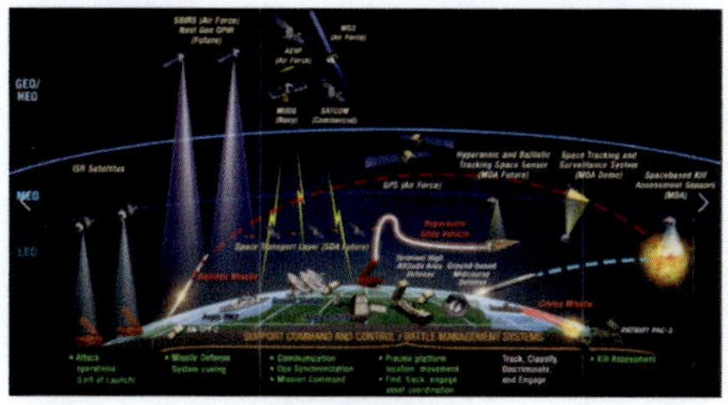

BMC2

SOURCE: BMC2 image reproduced from Jon Ahlbin, "MDA University Outreach QS Radiation Hardness Assurance for Parts and Materials (QSR) Division," briefing slides, Missile Defense Agency, 2021.
NOTE: BMC2 = Battle Management Command and Control; TRL = technology readiness level.

Weather as a service is an example of ready technology with the following dimensions:

- a single space-centered user community
- little integration with other systems.

At the opposite end of the complexity spectrum is BMC2, with the following dimensions:

- advanced technology requiring significant novel development
- a broad user community across domains and organizations
- extensive and critical integration challenges.

Other dimensions that lead to their own spectra of complexity include:

- stringent versus relaxed mission assurance requirements[11]

---

[11] Note that relaxed mission assurance requirements are scarce in space systems. In the past, such experiments as the Space Based Surveillance System, which started out as experiments with relaxed mission assurance requirements quickly became essential, leading to increased mission assurance requirements. See USSF, "Space Based Space Surveillance," fact sheet, March 22, 2017. Furthermore, while it is hypothesized that each satellite in a large fleet of satellites in low-earth orbit (where proliferation provides ample resilience) could be successful if built with relaxed mission assurance requirements, we note that common parts and systemic common failure mechanisms could devastate such a fleet. Diversity in design and manufacturing can mitigate the risk of common failures, but hardening to cyber, electronic, and radiation warfare will be essential even in a proliferated and diverse fleet. A real-world example is Russia's attacks on proliferated satellite ground terminals at the start of the 2022 war in Ukraine (Sandra Erwin, "U.S. Space Force to Step Up Protection of Satellite Ground Systems in the Wake of Russia's Cyber Attacks," *SpaceNews*, May 19, 2022c).

- small production runs of highly complex, atypical, and sensitive systems versus large production runs of standardized and nonsensitive systems.[12]

Given that the complexity varies in many dimensions for what the space community acquires, there is unlikely to be a one-size-fits-all solution to integration and synchronization challenges.

Approach

SAF/SQ requested that our research team recommend ways USSF can facilitate integration and synchronization of space acquisition and fielding to support the timely delivery of end-to-end space capabilities to the joint force. The research we initiated to address that request quickly became more generalized and expanded beyond integration, including any actions (not limited to just integration activities) that USSF could take that would lead to fielding more-timely, integrated, and synchronized warfighter capabilities. We explored the question from overlapping vantage points that include the stakeholder, user, and acquisition communities, as displayed in Figure 1.2.

Figure 1.2. Exploration of Overlapping Viewpoints

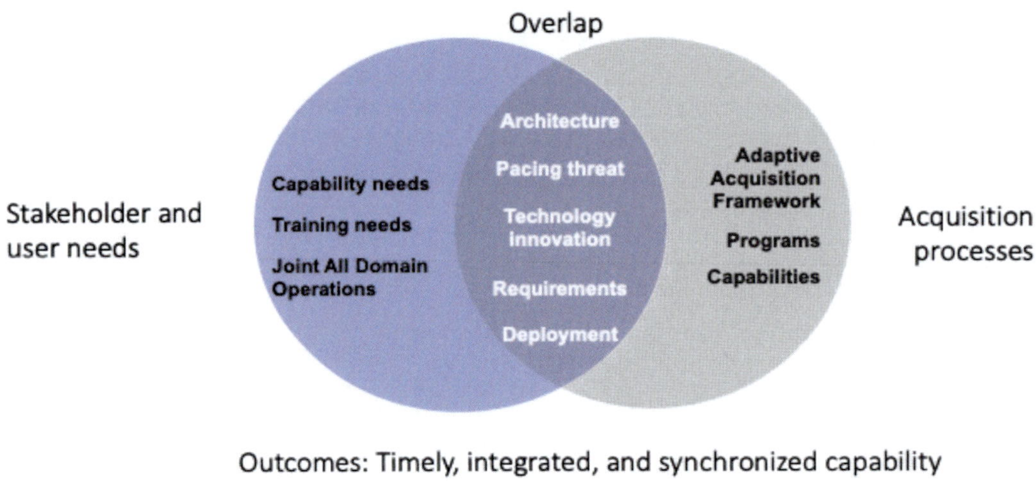

---

[12] Similarly, large production runs in space are scarce. The largest to date has been the SpaceX Starlink system, and because of the proprietary nature of that development, the complexity associated with managing a large and evolving production run has not been publicized (Jeffrey A. Drezner, *Competition and Innovation Under Complexity,* RAND Corporation, RP-1386, 2009).

Our methodology included a literature review, discussions with SMEs, and mapping the current state of the USSF acquisition ecosystem.[13]

Literature Review

Project AIR FORCE (PAF) assigned our research team to undertake a broad literature review of existing research on U.S government space-affiliated organizations to ensure that the analysis is leveraged and built on the key insights from this vast amount of research. Information was gathered from prior analyses conducted by federally funded research and development centers (FFRDCs), DoD (e.g., Defense Acquisition University and service-specific analyses), GAO, congressional reporting (e.g., Congressional Research Service and Congressional Budget Office), and trade literature relevant to options for improving integration and synchronization. We also compiled the latest DAF, USSF, and DoD-relevant policies to inform roles and responsibilities regarding synchronization and integration in space.

Discussions with Subject-Matter Experts

To supplement the literature review and to determine the current state of space acquisition, we conducted more than 60 semistructured interviews with SMEs and senior decisionmakers throughout DoD, other federal government agencies, and the private sector. Figure 1.3 lists some of the space community member organizations that were included and their roles (i.e., the reasons we chose to include these organizations). The overlapping perspectives were obtained through interviews with

- primary space system acquirers
- those who are responsible for requirements and oversight
- those who are responsible for organizing, training, and equipping the space community
- those who play additional supporting roles to the space community.

---

[13] We considered using case studies of program synchronization failures as part of our methodology. All programs that we considered as potential case studies failed differently and for a multitude of reasons. The integration and transition issues we found are systemic, not individual. Because the systemic issues are crosscutting, so are our methodology and our recommendations.

Figure 1.3. General Organizations and Roles of Interviewees

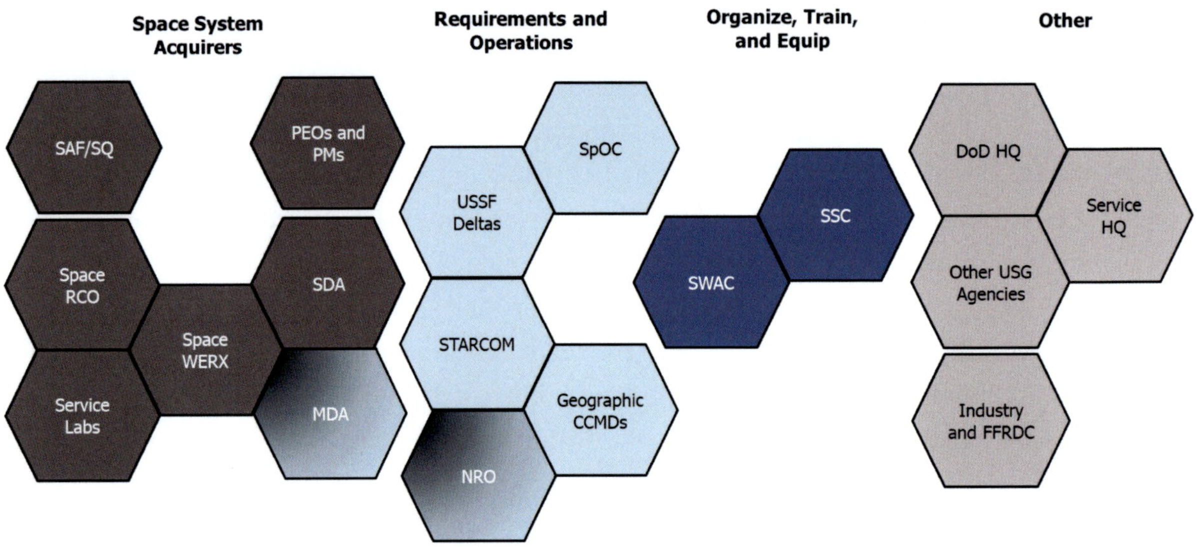

NOTE: CCMDs = combatant commands; FFRDC = Federally Funded Research and Development Center; HQ = Department of Defense headquarters; MDA = Missile Defense Agency; NRO = National Reconnaissance Office; PEOs = Program Executive Officers; PMs = Program Managers; SAF/SQ = Office of the Assistant Secretary of the Air Force for Space Acquisition and Integration; SDA = Space Development Agency; Service HQ = Service Headquarters; Space RCO = Space Rapid Capabilities Office; SpOC = Space Operations Command; SSC = Space Systems Command; STARCOM = Space Training and Readiness Command; SWAC = Space Warfighting Analysis Center; USG = U.S. government; USSF = United States Space Force.

Mapping of the Existing State of the United States Space Force Acquisition Ecosystem

The military space acquisition ecosystem has undergone significant change from 2020 to 2022, with the standup of USSF and congressionally mandated changes to space acquisition authorities. To understand the current state of the USSF acquisition ecosystem (e.g., the connections among the various stakeholders in space acquisition), we mapped a network of approximately 90 different stakeholders in USSF acquisition and the relationships among these organizations. We used the content from the stakeholder interviews and organizational change announcements to define the organizations.[14] We then graphed these existing relationships, as described in the interviews, using social network analysis tools to visually convey the results. It is important to note that we graphed the relationships for which we had evidence of being currently operational, not those that were called out in policy and doctrine. It is also important to note that this representation is current as of spring 2022. We acknowledge that these relationships are in flux as USSF matures. These graphs are first introduced in Chapter 2 and then used to highlight relevant relationships in later chapters. More specifics on the methodology are included in Appendix A.

[14] This approach allows a better understanding of how the connections exist in practice instead of how policies indicate connections should be.

Structure of This Report

In Chapter 2 of this report, we provide background on the organization and missions of USSF. In Chapter 3, we describe the need for overall space capability or mission roadmaps that could be analyzed for resilience against future budget volatility, a subject often cited in our interviews and the literature as a reason why it is difficult to deliver synchronized capability. In Chapter 4, we focus on the challenges in defining the capability or mission architectures within USSF. In Chapter 5, we discuss issues associated with aligning of the various space acquisition and operational organizations to enable better synchronization of developed and fielded capabilities. We discuss issues related to how to best harness the results of experimentation and innovation in space and how to transition new technology or capability into more-durable systems in Chapter 6. We focus on the transitioning of technology to operations, a perennial challenge, in Chapter 7. In Chapters 3–7, we provide our findings and recommendations at the beginning of each chapter, followed by more-detailed descriptions regarding each recommendation. Finally, we present our conclusions on addressing the issues and challenges we identify in Chapter 8. In addition, Appendix A provides additional details on the methodologies used in this research and Appendix B presents the interview protocols.

Chapter 2. Organization and Missions of the United States Space Force

Organization of the Department of Defense Space Acquisition Community

As we noted in Chapter 1, USSF is a new organization. Although USSF was established in December 2019, the Chief of Space Operations (CSO) was the only person assigned to USSF until a second person was assigned in April 2020. By summer 2021, more organizations had been stood up and at least nominally staffed. Shortly thereafter in October 2021, our team began the research that underpins this report, with the bulk of our interviews occurring in spring 2022. This report, therefore, should be viewed as a snapshot of the evolution of the force and additional work remains to be done in aligning the force.

Figure 2.1 shows the current organizations in the U.S. military space ecosystem and our findings regarding the strength of the relationships among them (see Appendix A for more details on our methodology). Qualitative data from the semistructured interviews were analyzed to develop this map of the stakeholders involved in space acquisition across USSF, DoD, IC, federal government, FFRDCs, and industry. The figure comprises approximately 90 nodes connected by 521 edges, with each node representing an organization, department, agency, or office. The nodes are color-coded as follows:

- organizations with formal acquisition authority (dark grey nodes)
- organizations with organize, train, and equip (OT&E) responsibility, but no formal acquisition authority (dark blue nodes)
- organizations in the operational community (light blue)
- all other organizations (light grey).

The size and font of the nodes represent the centrality of the organization within the acquisition ecosystem, computed as the weighted sum of the organization's interactions.[15] The figure shows the relationships between the organizations using connecting lines, and the width of those lines denotes the frequency of interaction. While frequency of interaction does not necessarily equate to quality, for this report, we use frequency as a reasonable proxy for the strength of the social interaction.[16]

---

[15] While it is common in social network analysis to define nodes with many strong connections (i.e., centrality) as having *influence*, we urge caution in using that word to describe these organizations. This is because influence in DoD acquisitions is perhaps more tightly tied to funding than to social interactions and may be more closely related to quality of interactions than to their frequency.

[16] It is important to note that the data used in this figure are an output of the over 60 semistructured interviews. The connections and frequency of interactions result from those interviews and reflect the relevance provided by the interviewees.

Figure 2.1. Space Acquisition Stakeholder Map

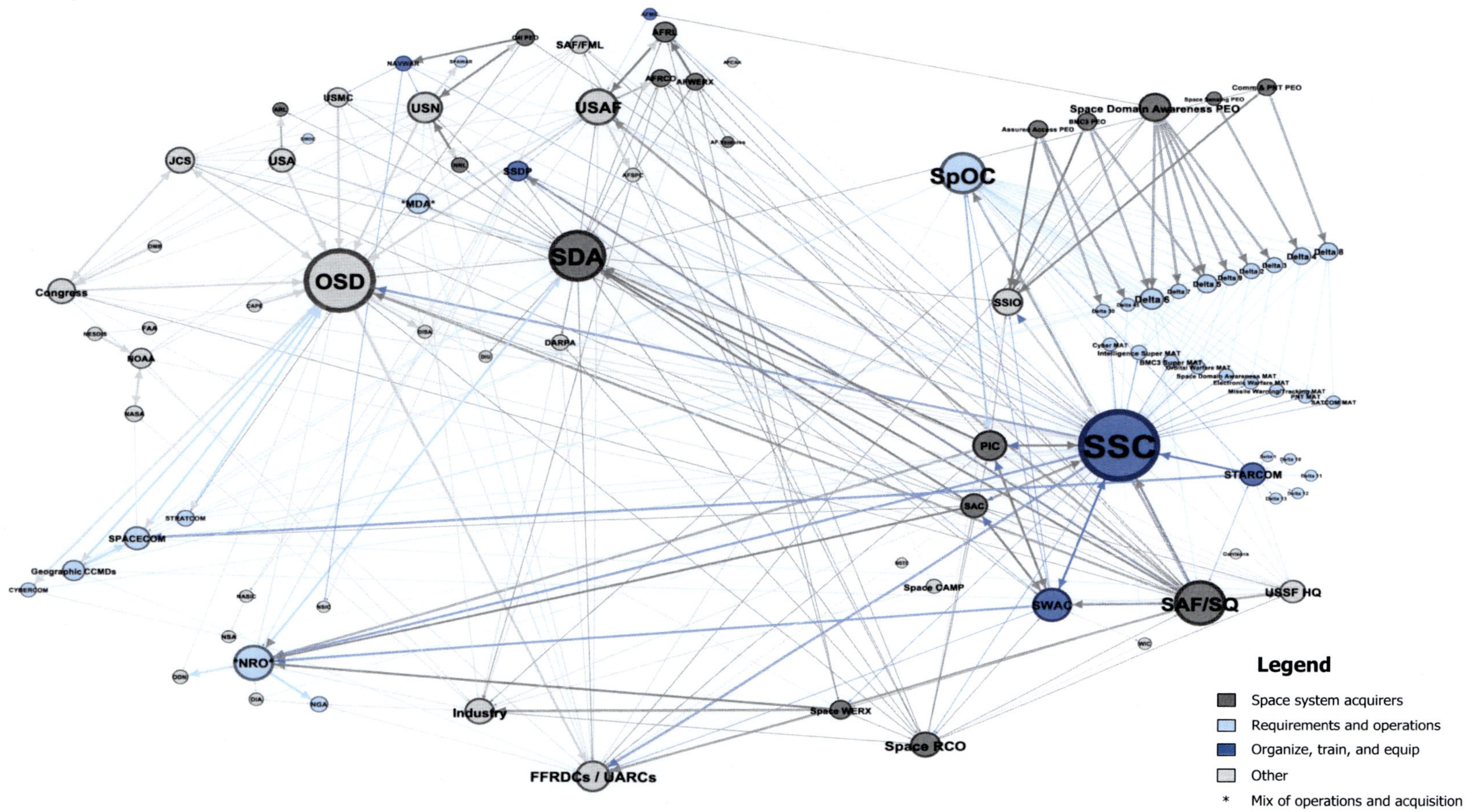

SOURCE: Features RAND analysis of interviews conducted in spring 2022.
NOTE: Node and font size indicates the centrality of the organization, calculated as the weighted sum of its interfaces. Line width indicates the frequency of the relationship as recorded in the interviews. This diagram depicts organizational interactions on the basis of interviews. Therefore, the diagram reflects the as-is state of the ecosystem in spring 2022, prior to the SAE position being filled.

We laid out the map by loosely grouping the organizations that make up USSF (on the right), other DoD organizations (left of center top), other U.S. government organizations and, finally, other organizations (left of center bottom). Therefore, spatial proximity on the map is determined by organization, not by social proximity or influence.

Table 2.1 summarizes some of the significant organizational changes that have occurred from December 2017 to May 2022 because DoD has adjusted acquisition accountabilities to respond to the emergence of space as a warfighting domain. Key acquisition organizations are highlighted in the next subsections.

Table 2.1. Sample of Recently Created Military Space Organizations

Organization	Role	Date of Creation or Designation
Space Rapid Capabilities Office	Acquisition (rapid)	December 2017
Space Development Agency	Acquisition (disruptive)	March 2019
U.S. Space Command[a]	Combatant command	August 2019
U.S. Space Force[b]	Military service	December 2019
Office of the Assistant Secretary of the Air Force for Space Acquisition and Integration	Acquisition (oversight)	December 2019
Space Acquisition Council	Acquisition (oversight)	December 2019
Space Operations Command	Operations	October 2020
Program Integration Council	Acquisition (coordinate)	September 2020
Space Warfighting Analysis Center	Requirements	April 2021
Space Systems Command	Acquisitions (all)	August 2021
Space Training and Readiness Command	Training	August 2021
SpaceWERX	Acquisition (science and technology)	August 2021
USSF Service Acquisition Executive[d]	Acquisition (oversight)	May 2022

[a] For organizations that share names with previous entities (e.g., U.S. Space Command), we list the creation date for the current incarnation.

[b] Although USSF was created in December 2019, it had only one member until a second was added in April 2020. The transfer of U.S. Air Force personnel then slowly ramped up and was largely complete by summer 2021. Transfers from other services did not occur until summer 2022.

[c] The lineage of Space Systems Command, formerly the Space and Missile Systems Center, can be traced back to the Space Systems Division of the Air Research and Development Command established in 1961. Space Systems Command has long been a dominant player in space acquisition. Today, the Space Systems Command commander has no Program Executive Officer role, but Space Systems Command remains the home for five of the space Program Executive Officers within DoD.

[d] Prior to the USSF Service Acquisition Executive confirmation by the Senate, the Office of the Assistant Secretary of the Air Force for space performed the acquisition executive role for USSF.

Key Acquisition Organizations

Space Rapid Capabilities Office

Space RCO was created by Congress in December 2017. Its mission, as specified in U.S. law, is to focus on low-cost, rapid reaction capabilities.[18] Space RCO's focus on rapid reaction to threats requires prioritizing the use of existing technology and getting capability into operations quickly. Space RCO is also exempt from the Joint Capabilities Integration and Development System (JCIDS).[19] Space RCO's requirements and acquisition priorities are determined by the organization's board of directors, which is chaired by the Secretary of the Air Force and includes the CSO and the Chief of Staff of the Air Force.[20]

Space RCO's first project started in 2019. All projects currently in development at Space RCO are classified and very little is publicly known about them. However, the organization did announce that it would deliver its first project into operations in 2022.[21] Space RCO has a dedicated PEO and, because of its small size, limited number of programs, and urgent mission, is able to staff its program management and contracting teams with highly experienced individuals. In our interviews, Space RCO personnel noted that this ability to staff with highly experienced personnel is one of the keys to Space RCO's success.

Space Development Agency

SDA was created by Congress in 2019 as a separate DoD agency and transitioned to USSF in 2022. Its mission is to develop and demonstrate innovative satellite systems to support capabilities in "transport, battle management, tracking, custody, navigation, deterrence, and support," often through the procurement of "commercial capabilities and services."[22] SDA is currently committed to demonstrating capabilities in two-year increments, called tranches, with the first due in 2022 and the second in 2024. Tranche 1 was focused on the satellite communications mission, which SDA calls its transport layer.[23] According to the U.S. Code, the Secretary of the Air Force and the USSF SAE are to ensure SDA is delegated contracting

---

[18] U.S. Code, Title 10, Section 2273a, Space Rapid Capabilities Office.

[19] U.S. Code, Title 10, Section 2273a.

[20] Space Rapid Capabilities Office, "Space Rapid Capabilities Office (Space RCO) Overview," briefing slides, October 14, 2020, slides 3 and 5.

[21] Courtney Albon, "US Space Force Rapid Capabilities Office to Deliver First Project This Year," *C4ISRNET*, July 7, 2022b.

[22] Jon Ludwigson, *Space Acquisitions: DoD Faces Challenges and Opportunities with Acquiring Space Systems in a Changing Environment*, testimony before the Subcommittee on Strategic Forces, Committee on Armed Services, House of Representatives, Government Accountability Office, GAO-21-520T, May 24, 2021a, p. 3; U.S. Code, Title 10, Section 9086, Space Development Agency.

[23] DoD, "Space Development Agency Makes Award for 126 Satellites to Build Tranche 1 Transport Layer," press release, February 28, 2022a.

authority and milestone decision authority for MTA programs.[24] Like Space RCO, SDA has a dedicated PEO and is selective when staffing its contracting and program management workforce.

Space Systems Command

The dominant space acquisition organization since 1961, the Air Force's Space and Missile Systems Center (SMC) was reorganized in 2018. It subsequently transitioned to USSF and was renamed Space Systems Command with some changes in roles and responsibilities. At the direction of the Secretary of the Air Force, SSC reorganized again in 2021. In its latest incarnation, SSC is the second USSF field command and, although it houses the majority of USSF's acquisition personnel (and five of the seven USSF PEOs), the SSC commander is not in the acquisition chain of command. While the SMC commander had both acquisition and OT&E authorities, as of this writing, the SSC commander has only OT&E authority, and acquisition authority resides with the PEOs within SSC.[25] It is worth noting, per a SSC realignment briefing from December 2021, that SSC views its role as "space systems-of-systems (SoS) integrator, advisor to the SAE at all space milestone decisions, [and] budget integration/execution authority responsible for programming resources."[26] However, this role has not been formally described or documented at the time of this report's writing (summer 2022).

Other Significant Changes

Some legacy acquisition programs have only recently (in summer 2022) been transferred from other parts of DoD to USSF, most notably for the satellite communications mission, and former Navy and Army satellite communications operators and planners have been transferred to USSF. Also, STARCOM, the third USSF command, was formed in late 2021 to improve USSF's testing and training infrastructure.

On the operations side, the heritage Space Operations Squadrons and Space Warning Squadrons are now organized into deltas within SpOC, and new deltas with a focus on electromagnetic warfare and better integrated intelligence have been stood up. Major workforce transitions and hiring have also been happening within USSF, along with discussions on roles and responsibilities for the various leadership positions and accompanying offices.

---

[24] U.S. Code, Title 10, Section 9086.

[25] The PEOs within SSC rely on the SSC commander for functional support and report to the SAE, from whom they derive their acquisition authority. For a good tutorial on current command relationships, see Theresa Hitchens, "Rocket Science: How Space Force Acquisition Works, with Many Players and Dual Hats," *Breaking Defense*, July 27, 2022c.

[26] Space Systems Command, "SSC Realignment," United States Space Force, briefing slides, version 14.0, December 8, 2021.

These rapid changes in the space acquisition ecosystem mean that it is too soon to judge these changes' effectiveness in synchronizing and integrating space systems. To the extent that prior synchronization and integration issues were a product of benign neglect, a new emphasis and focus can help. However, if roles and responsibilities are overlapping or are not supportive of the production of an integrated architecture that is responsive to operational need, these changes might make an already complex situation worse. We will return to this topic in Chapter 4, "Architectural Focus."

Missions of the United States Space Force

As we will discuss in more detail in Chapter 4, there are many ways to partition USSF missions into capability portfolios and SoS architectures. For the purposes of this report, we partition the missions as[27]

- weather
- position, navigation, and timing (PNT)
- satellite communications
- missile warning and missile defense
- space domain awareness
- protecting and defending.

Below, we discuss each mission and highlight important innovations, ongoing acquisition activities, and past synchronization failures.

Weather

USSF weather missions have traditionally been separated into terrestrial weather sensing and space weather sensing.

Today, DoD's terrestrial weather sensing capabilities are largely provided by the National Oceanic and Atmospheric Administration (NOAA), and a large number of different U.S. and

---

[27] Intelligence, reconnaissance, and surveillance (ISR), nuclear detonation detection, and space launch are not in our list of missions. Our reasoning for this is that

- the space-based ISR mission is currently the responsibility of NRO and the National Geospatial-Intelligence Agency, and the air-based ISR mission is currently the responsibility of the U.S. Air Force (USAF). While some in USSF leadership are interested in taking on responsibility for a more tactically oriented space-based ISR mission, the USSF SAE has stated that they will continue to look to NRO and the National Geospatial-Intelligence Agency to fill that role, and the Secretary of the Air Force has indicated that a closer relationship will be sought with NRO (Theresa Hitchens, "NRO, Air Force May Co-Fund Future Space-Based ISR Tech: Kendall," *Breaking Defense*, August 1, 2022d).
- the nuclear detonation detection mission is the responsibility of the National Nuclear Security Administration, although USSF-operated satellites do host sensors needed for that mission.
- space launch is a service in support of USSF missions, not a mission itself. Space launch is now acquired as a service from commercial entities. USSF acquires those services in support of both USSF and NRO missions.

allied civil and commercial satellites contribute to this mission. In addition to operating its own weather satellites, NOAA also operates the remaining satellites of a 1960s-era DoD unique weather sensing constellation—the Defense Meteorological Satellite Program (DMSP).[28] DMSP operations are performed in collaboration with a small USSF detachment at the NOAA operations center. The future of unique DoD terrestrial weather sensing satellites is uncertain. While studies are underway to evaluate the acquisition of new systems, there is also discussion about DoD simply acquiring terrestrial weather products "as a service."[29] Given that DoD has not acquired a new terrestrial weather system since the 1970s, there is no recent evidence of integration or synchronization issues. However, any new system will need to be integrated not just with NOAA but also with the weather data processing and distribution centers at the USAF Weather Agency and the U.S. Navy's Fleet Numerical Meteorology and Oceanography Center. Synchronization across the services has historically been difficult.

Space (as opposed to terrestrial) weather monitoring includes detecting solar flares and measuring natural radiation levels in space, among its primary functions. Both DoD and NOAA host small sensors on board a variety of military, civil, and commercial satellites to monitor the weather of space. While measuring natural radiation levels can only be done in space, monitoring solar activity can be done from either the ground or space. The acquisition and fielding of these sensors has been an area of recent experimentation and innovation.[30] Space weather information is a critical element of space domain awareness, and the importance of the space weather monitoring mission will likely grow if conflicts extend more actively into the space domain.

Position, Navigation, and Timing

The U.S. Global Positioning System (GPS) was the first space-based system to provide high-precision position and time of day ubiquitously across the globe. This capability has revolutionized not only warfighting but also daily life for hundreds of millions of civilians. Today, GPS is a global utility. Many countries have also developed their own PNT systems,

---

[28] NOAA, "Defense Meteorological Satellite Program (DMSP)," Office of Satellite and Product Operations, webpage, undated-a.

[29] USSF has plans to build two satellites to meet DoD's unique requirements. Three companies were selected in 2020, but these programs have faced continual funding challenges (Sandra Erwin and Brian Berger, "A Race Against Time to Replace Aging Military Weather Satellites," *SpaceNews*, May 24, 2021; Courtney Albon, "US Space Force Considers Purchasing Weather Data as a Service," *DefenseNews*, January 11, 2022a).

[30] See, for instance, the award of two contracts related to space weather sensing in the article by Debra Werner, "Air Force Taps Applied Technology Associates and Teledyne Brown to Produce Space Weather Sensors," *SpaceNews*, January 29, 2020. A less traditional approach to monitoring space weather using information collected by ham radio operators is described in K. Collins, D. Kazdan, and N. A. Frissell, "Ham Radio Forms a Planet-Sized Space Weather Sensor Network," *Eos*, February 9, 2021.

including Russia's GLONASS system, China's BeiDou system, the European Space Agency's Galileo system, and Japan's Quasi-Zenith Satellite System.[31]

U.S. allies' PNT systems are acquired and operated by civilian government organizations and do not have signals specifically designated for military PNT. However, the Russian, Chinese, and U.S. systems are more explicitly designed to support warfighting operations. Although then-President Bill Clinton removed restrictions that reserved higher-accuracy GPS signals for military use in 2000, DoD has since implemented a specific military code (M-code) on later GPS satellites.[32] M-code is not yet operational because it became desynchronized from the ground segment and user equipment needed to complete the capability. Delays in making M-code operational across the armed services is a well-known illustration of the lack of both vertical (e.g., space to ground) and horizontal (e.g., USSF to other U.S. armed services) synchronized fielding of DoD space systems and those delays have been well-documented.[33]

Satellite Communications

The mission to provide satellite-based communications services to DoD is often separated by whether that mission is supporting tactical operations, such as those between a joint task force commander and a forward operating element, or strategic operations, such as nuclear command, control, and communications (NC3).[34] However, DoD also relies heavily on commercial satellite communications systems for tactical operations, which it purchases as a service from a wide variety of vendors in collaboration with the Defense Information Systems Agency. Given the wide variety of interfaces and operational contexts within which these disparate systems were designed, it is a daunting task to integrate and synchronize across them. An additional

---

[31] These systems are not interoperable with GPS in that they use different downlink frequencies and absolute timing references. However, many of today's commercial and civilian receivers can switch between systems, providing for a seamless user experience.

[32] Reuters, "Clinton Acts to Make GPS More Accurate," *New York Times*, May 2, 2000.

[33] As a 2022 GAO report notes, "The Department of Defense (DoD) has been modernizing GPS to use a more jam-resistant, military-specific signal called M-code for more than two decades. The Space Force is nearing a key GPS milestone—a constellation of 24 satellites broadcasting the signal. However, while the ground segment that controls the signal will be available in 2023, widespread operational use will take several more years, due to developmental challenges and delays with the user equipment that will receive the signal" (Jon Ludwigson, *GPS Modernization: Better Information and Detailed Test Plans Needed for Timely Fielding of Military User Equipment*, Government Accountability Office, GAO-22-105086, May 9, 2022).

[34] Given that strategic communications satellites are increasingly used to support tactical operations, it has become common in recent years to separate USSF satellite communications systems as being either *protected* or *unprotected* against adversarial electronic warfare, although those words are problematic given that no USSF communications satellite is truly unprotected against adversarial electronic warfare. For this report, we will use *tactical* when the core mission of a satellite constellation is to support theater operations and *strategic* when its core mission is to support NC3. Another common partitioning of military satellite communications is to distinguish between narrow-band and wide-band communications systems. In this case, *wide-band* refers to satellites that can transmit in a wide range of frequencies and *narrow-band* means that transmission is within the coherence band of a frequency channel ("Narrowband vs. Wideband," *JEM Engineering* blog, September 2021).

synchronization problem is that the handsets, terminals, and gateways used to connect to the satellite communications systems (i.e., the user equipment) are acquired by the using services, not USSF.

In one of his first acts after the USSF's standup in January 2020, then–Chief of Space Operations Gen John W. Raymond signed a white paper, *United States Space Force Vision for Satellite Communications (SATCOM)*, which outlined some of the challenges in synchronizing satellite communications across DoD and his vision for the future.[35] Innovations in satellite communications acquisition have centered on finding creative ways to partner with other countries and on creating user equipment that can "roam" from one system to another to reduce the number of user equipment suites that need to be deployed into operations.[36] Another experiment has focused on implementing space-based detection of spectral interference and autonomously routing communication between satellites to avoid the interference.[37] For experimentation on a much larger scale, in early 2022, SDA awarded contracts for Tranche 1 of an envisioned network of 126 small communication satellites in low-earth orbit interconnected with optical links to provide a "resilient, low-latency, high-volume data transport communication system."[38]

Missile Warning and Missile Defense

The traditional mission for space-based missile warning is tied to U.S. Strategic Command's homeland missile defense mission, which uses a network of both space- and ground-based sensors to detect possible incoming threats. The ground sensor locations and look angles are designed to sense incoming ballistic missiles following relatively fixed trajectories from specific launch sites. In response to the threat of hypersonic missiles, which have a much wider range of possible approach trajectories, future U.S. strategic missile warning architectures appear to emphasize a global space-based sensor network with little to no augmentation from ground-based sensors.[39]

---

[35] DAF, *United States Space Force Vision for Satellite Communications (SATCOM)*, January 23, 2020.

[36] Consider, for example, the placement of a hosted military satellite communications payload on a Norwegian satellite, detailed by Sandra Erwin, "Space Force Delivers First of Two U.S. Payloads to Launch on Space Norway's Arctic Broadband Mission," *SpaceNews*, June 9, 2022d. These innovations have included several pilot programs under the Flexible Modem Interface initiative to demonstrate "that a terminal or network can be reconfigured in response to interference, congestion, or changing mission needs in a matter of minutes as opposed to weeks or months, as is currently the case" (Larry Simon, "EM&C and FMI—Taking Small Steps to an Integrated Satellite Architecture," Government Satellite Report, September 10, 2020).

[37] Theresa Hitchens, "Newest Sats Launched by DoD Include Jammer-Evading, Classified Payloads," *Breaking Defense*, July 6, 2022a.

[38] DoD, 2022a.

[39] We write this statement with some caution because it is based on our review of recent DoD budget requests, including those from the MDA, SDA, and USSF, and congressional allocations for this mission to those entities. However, given our finding regarding the absence of definitive architectural roadmaps discussed later in this report, the shift we perceive may not be deliberate or permanent.

More recently, space-based missile warning has become increasingly important for theater (or tactical) missile defense. For theater missile defense, DoD combines space-based sensing of missile launch with ground radars positioned in or near the theater of operations to track the missile trajectory through midcourse and terminal approach. Data from these sensors are then fed to a variety of interceptors.

Interoperability standards are key to synchronizing fielding of the SoS that make up missile warning and missile defense. Currently, DoD's JADC2 effort has a stated goal of developing interoperability standards to connect "any sensor to any shooter," which would apply to both strategic and theater missile defense. However, the release of these standards is still some years away.[40] For the specific case of strategic missile defense, MDA has developed an integrated battle management system to broker the interfaces between the variety of sensors and interceptors used.[41]

The missile warning mission area is the source of many USSF experiments and innovations that explore everything from new types of sensors to new acquisition approaches. Among the sensor experiments, a wide field-of-view infrared staring sensor was launched in June 2022 into geosynchronous orbit with the goal of providing vastly improved tracking of hypersonic glide vehicles, which have a relatively dim infrared signature.[42] Experiments in acquisition approaches include those used by SDA to acquire a space tracking layer of small satellites in low-earth orbit to complement its space communications transport layer. In the meantime, SSC is conducting experiments and studies for a "next generation overhead persistent infrared sensing system" and MDA is experimenting with a concept for a "hypersonic and ballistic tracking space sensor."[43] As we noted earlier, interoperability standards will be key to ensuring that the results of this experimentation can be harnessed into a fielded SoS. Unfortunately, there are already warning signs that interoperability is being deprioritized.[44]

---

[40] Our statement regarding the probable release of standards for this purpose is based on a current multiyear RAND project for the Office of the Secretary of Defense (OSD) Research and Engineering developing standards to support the JADC2 effort.

[41] This statement is based on 2019 work RAND did in support of the JADC2 initiative, which examined the MDA battle management system in some depth. We are not aware of similar efforts to integrate the tactical missile warning and missile defense missions (James Dimarogonas, Jasmin Léveillé, Jan Osburg, Shane Tierney, Bonnie L. Triezenberg, Graham Andrews, Bryce Downing, Muharrem Mane, and Monica Rico, *Universal Command and Control Language Early System Engineering: Performance Effects of a Universal Command and Control Standard*, RAND Corporation, RR-A744-2, 2023).

[42] Amanda Miller, "Experimental IR Satellite Heads to GEO to Advance Hypersonic Missile Warning and Track," *Air and Space Forces Magazine*, June 29, 2022.

[43] Greg Hadley, "Study: Combine Missile Warning, Tracking Constellations into One Multi-Orbit System," *Air and Space Forces Magazine*, June 7, 2022.

[44] John D. Sawyer, *Missile Defense: Better Oversight and Coordination Needed for Counter-Hypersonic Development*, Government Accountability Office, GAO-22-105075, June 16, 2022.

Space Domain Awareness

The term *space domain awareness* covers a range of activities that we will loosely categorize as surveillance, situational awareness, and domain awareness. Many U.S. allies and commercial entities contribute to space *surveillance* to observe and record the actions of objects in space.[45] *Situational awareness* requires the ability to systematically capture, record, validate, and share the data collected through surveillance. For DoD, the sharing of space situational awareness data is handled by the USSF 18th and 19th Space Defense Squadrons.[46] At the highest levels, *domain awareness* requires the ability to analyze data and predict and synthesize facts and evidence into an assessment of possible threats and probable future outcomes. Threat assessment of object behavior is highly intertwined with the U.S. intelligence community.

Protect and Defend Space Assets

While "defining resilient space order of battle and architectures (defensive and offensive)" is a top operational imperative of the Secretary of the Air Force, there is no currently published, authoritative architectural roadmap for evolving existing USSF systems to a coordinated space battle management system.[47]

At the core of any space battle management capability will be a system for tracking and cataloging the millions of objects in space. There have been several attempts over the decades to upgrade existing systems to better form an integrated SoS architecture for space battle management. A 2019 GAO report cites three notable attempts: the Cheyenne Mountain Upgrade (1981–1998, 11 years late, $1 billion over budget, with key capabilities undelivered), the Combatant Commander's Integrated C2 System (2000–2006, three years late, and $250 million over budget when terminated with key capabilities undelivered), and the Joint Space Operations Center Mission System (2009–2018, three years late, and $200 million over budget when terminated with key capabilities undelivered).[48]

Both organizational and technical issues have been cited as reasons for failure. The primary organizational issue is that many of the systems to be integrated have been outside the control of

---

[45] As an example of such collaboration, in 2020, the United States and Japan signed an agreement under which the United States acquired space domain awareness optical sensors to be flown on Japanese PNT satellites (Space Force Public Affairs, "Japan's Office of National Space Policy Signs Historic MOU with the U.S. Space Force," Space Force News, December 18, 2020).

[46] For recent innovation in the sharing of space situational awareness, see Blythe Goya, "18th Space Control Squadron recognized for innovation," Vandenberg Space Force Base, Combined Force Space Component Command Public Affairs, November 29, 2021; this details a team nominated for the 2021 Col Bradford Parkinson Innovation Award.

[47] Charles Pope, "Kendall Details 'Seven Operational Imperatives' and How They Forge the Future Force," Secretary of the Air Force Public Affairs, March 3, 2022.

[48] Cristina T. Chaplain, *Space Command and Control: Comprehensive Planning and Oversight Could Help DoD Acquire Critical Capabilities and Address Challenges*, Government Accountability Office, GAO-20-146, October 30, 2019b.

the implementing office. This has been largely corrected with the standup of USSF. Technical issues have centered on accuracy, reliability under load (e.g., scalability), and security.[49] These technical issues *only* can be solved, in our opinion and in that of others who have studied this issue, within an overall architecture and modernized computing infrastructure.[50] However, USSF has yet to put forward an integrated architecture, instead favoring an iterative approach to fielding new applications on top of older computing infrastructure to meet immediate needs.

A new infrastructure (or backbone) for space battle management, called the Advanced Tracking and Launch Analysis System (ATLAS), is in development at SSC. SDA is also conducting experiments related to space-based battle management command, control, and communication (BMC3) through a battle management layer.[51] It is not entirely clear what the scope of the BMC3 effort is, how it might interact with ATLAS, or its relationship to a previous SSC-led effort, titled Enterprise Battle Management Command and Control.[52] Another entrant into space battle management design is the Future Operationally Resilient Ground Evolution (FORGE) system, which is being designed to support missile warning and missile defense satellite operations.[53] Although limited to a single mission, FORGE has significant overlap with the overall space battle management function that must be integrated and synchronized.

In addition to the integrating battle management function, well over 100 programs within USSF are experimenting with and/or developing offensive and defensive capabilities to protect and defend U.S. and allied space systems. A common partitioning of these programs is by the type of warfare they support: cyber, electromagnetic, or kinetic.[54] While most programs are classified, unclassified examples in the electromagnetic domain include the Bounty Hunter system, designed to provide situational awareness of spectrum interference, and the Counter Communication System, a transportable jamming system.[55] Details of many of these programs

---

[49] As early as 2012, DoD's Developmental Test and Evaluation organization was actively working issues "of system performance, reliability, interoperability, and information security" in Joint Mission System Increment 2 (DoD, *Developmental Test and Evaluation: FY 2012 Annual Report*, March 2013).

[50] In 2019, an unpublished RAND study illuminated the challenges DoD could expect to encounter when implementing space situational awareness and battle management. Some of the findings of this report regarding the challenges of integration were raised in that study.

[51] SDA, "Battle Management Command, Control, and Communication (BMC3)," webpage, undated.

[52] Theresa Hitchens, "New Space BMC2 System to Debut at Corona: Raymond," *Breaking Defense*, September 30, 2019.

[53] While the backbone of FORGE is being developed by Raytheon, several small mission application prototyping efforts have been awarded to smaller firms. See, for instance, "Altamira Awarded an $8 Million+ Space Mission Data Processing Award by USSF," *SatNews*, December 10, 2020.

[54] Another common partitioning is between offensive space control and defensive space control, but use of these terms appears to be waning.

[55] Frank Wolfe, "U.S. Space Force Unfunded Priorities List Includes $431 Million for Programs to Provide 'Warfighting Punch,'" *Defense Daily*, June 7, 2021.

are classified and compartmentalized within different Special Access Programs, complicating efforts to synchronize acquisitions even at the very highest levels of USSF.[56]

[56] According to DoDD 5205.07, a Special Access Program is established to safeguard information by imposing access requirements that "exceed those normally required for information at the same classification level" (DoDD 5205.07, *Special Access Program (SAP) Policy*, U.S. Department of Defense, July 1, 2010, incorporating Change 2, February 4, 2020).

Chapter 3. Budget Resiliency

Even though the DoD Planning, Programming, Budgeting, and Execution (PPBE) process attempts to provide budget stability by projecting needs over a five-year period, and despite relatively stable top-line congressional defense appropriations, research has found that "budget outcomes are frequently volatile and unpredictable at the individual program level."[57] Budget instability at the program level and budget inflexibility overall make it difficult to execute and deliver integrated and synchronized capabilities. The resilience of program planning to budget instability has not historically been a consideration when partitioning mission capability between USSF program elements (PEs), and there has been insufficient objective analysis of the impacts of budget instability on the space industrial base.[58] In addition, the relative inflexibility of DoD's PPBE system makes it difficult to field capabilities to counter rapidly evolving threats. Budget uncertainty—and the disruptions it brings—is a common theme both in the literature and in our analysis of interviewee responses regarding the difficulties of synchronizing the fielding of integrated space capabilities.

Budget volatility has worsened in recent years; there has been an increased reliance on continuing resolutions in years when Congress has not appropriated funding prior to the start of the fiscal year. Continuing resolutions have a particularly large impact on transitioning technology from experimentation to programs of record (PoRs) because new programs or programs that are reliant on a ramp-up in funding cannot be funded.[59] In a 2021 report examining the impact of continuing resolutions on program execution, GAO noted that it is now routine practice within DoD to formulate spending plans assuming 30-, 60-, and 90-day continuing resolution periods at the start of each fiscal year to mitigate the impact of the continuing

---

[57] Heidi Brockmann Demarest, *US Defense Budget Outcomes: Volatility and Predictability in Army Weapons Funding*, Palgrave Macmillan, 2017, p. 2.

[58] The PE is the primary data element and foundation of the PPBE process. It is the unit to which resources are allocated.

[59] There is an anomaly request process by which DoD can ask for an exception to this prohibition against new starts or increased funding, but few anomaly requests are successful. For example,

> Congress granted only one, allowing the Navy to enter into a contract and use incremental funding for the procurement of two Columbia-class submarines, in FY [fiscal year] 2021. The final CR [continuing resolution] did not authorize other requested anomalies, including the new start for the W93 warhead program, reallocation of funding for weapons activities in the National Nuclear Security Administration, *and the transfer of funding from the Air Force to the Space Force* . . . [emphasis added] (Seamus P. Daniels and Kaitlyn Johnson, "What the Continuing Resolution Means for Defense and Space Force Funding in FY 2021," Center for Strategic and International Studies, October 9, 2020).

resolutions.[60] Although GAO found that the impact of continuing resolutions was mitigated for the established programs that it examined, the report did not examine the impact of continuing resolutions from an entire portfolio perspective, from which the impact of delayed program starts or delayed program funding increases may have been more apparent.

While working more closely with Congress and other stakeholders in the financial management community may reduce some budget uncertainty in space acquisition, USSF will likely need to adopt tools and techniques that allow it to manage the resilience of its investment planning to that budget uncertainty. This way, USSF can create a virtuous cycle, in which improved resilience to budget uncertainty yields better acquisition and warfighting outcomes, which then improves USSF's relationship with Congress, and reduces budget uncertainty. Figure 3.1 depicts this progression, and Table 3.1 summarizes our findings and recommendations related to budget resilience. The remainder of this chapter describes these findings and recommendations more fully.

Figure 3.1. The Virtuous Cycle of Budget Resilience

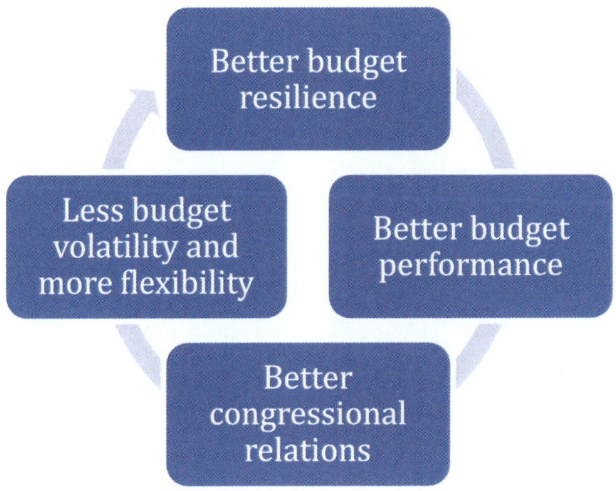

[60] Elizabeth A. Field, *Defense Budget: DoD Has Adopted Practices to Manage Within the Constraints of Continuing Resolutions*, Government Accountability Office, GAO 21-541, September 13, 2021.

Table 3.1. Findings and Recommendations Related to Budget Resilience

Findings	Recommendations
Budget instability at the program level, and budget inflexibility overall, make it difficult to execute and deliver integrated and synchronized capabilities. • Budget instability resilience has not historically been a consideration when partitioning mission capability between USSF program elements.[a] • The relative inflexibility of DoD's Planning, Programming, Budgeting, and Execution system makes it difficult to field capabilities to counter rapidly evolving threats. • There is insufficient understanding of economic interactions between USSF and the broader space market and how budget instability affects the space defense industrial base.	Create resource-loaded roadmaps of mission capability architectures and conduct sensitivity analyses to better understand and prepare for the effects of budget actions, such as continuing resolutions, changes to funding levels, or program cancellations. • Use the outputs of the analyses to enable better communication of mission capability trade-offs related to funding choices. • To mitigate the impacts of budget instability, incorporate modularity into the budgetary design of program elements and capability roadmaps. • To mitigate the impacts of budget inflexibility, create program elements[a] that aggregate planning items to cover a significant portion of a mission capability.[b] • Develop tools, processes, and methods to assess (1) the resilience of USSF capability roadmaps to changes in the industrial base and (2) the impact of USSF acquisition strategies and/or budgeting decisions on the health of overall market economy that sustains the base.
More-transparent Congressional interaction will be needed if USSF is to mitigate the impact and magnitude of budget instability.	Short term: Implement a cultural change that values more-open communication with Congress and builds alignment across USSF (field commands and acquisition community), Department of the Air Force, the Pentagon, and the greater U.S. space enterprise. • Continue to improve USSF workforce education on budget processes involving the larger (DoD, White House, and Congress) financial management community. • Ensure that USSF personnel understand and present acquisition strategies and plans consistently to all external stakeholders. Long term: As USSF budget performance improves and matures, request that Congress increase the threshold for reprogramming authority to allow the Secretary of the Air Force to more easily reprogram funds between program elements in a mission capability, with advice from the Chief of Space Operations and Service Acquisition Executive.

[a] The program element is the primary data element and foundation of the Planning, Programming, Budgeting, and Execution process. It is the unit to which resources are allocated.
[b] For the protect and defend mission, it is challenging to create aggregated program elements because very few people are cleared to the right levels to understand all technologies in that portfolio.

Conducting Sensitivity Analyses to Budget Uncertainty

To fully understand the effects of budget uncertainty on a portfolio, USSF should develop capability roadmaps (also discussed in Chapter 4) that show all of the prototypes, experiments, rapid capabilities, and established PoRs that currently or potentially might contribute to that capability. These roadmaps would have time-sequencing information to allow a portfolio manager to develop a plot of net capability (preferably expressed as a mission effectiveness metric) over time. As with any integrated schedule, the roadmap should be resource-loaded.[61] If the resources are expressed as a function of funding and if start and end dates are allowed to float in response to funding availability, then the roadmap can be used to conduct sensitivity analyses on the impact of factors (such as budget delays, changes in funding level, cancelled or failed prototyping, or program efforts) on mission effectiveness. From these sensitivity analyses, the portfolio's resilience, or lack thereof, to budgetary uncertainty can be demonstrated.[62]

Creating Resilient Investment Plans

However, it is not enough simply to measure the budgetary resilience of current portfolios. Improving budgetary resilience requires that the knowledge gained from the sensitivity analyses be incorporated into investment planning. Given that budgets are both unstable (as previously discussed) and inflexible (as will be discussed), designing for resilience requires both modularizing activities to protect against instability and aggregating activities to allow for more agility and freedom of action between budget cycles. Although modularity is often associated with dividing a plan into smaller parts and aggregation is associated with assembling items into larger parts, these concepts are not opposites—in an ideal world, USSF would aggregate modular parts into its investment planning. For example, a modular program planning element would aggregate all space and ground elements needed to implement the GPS M-code capability into a single PE. The funding and capability delivery dates would then move together in response to shifting congressional priorities, but the SAE would have the freedom to move money among the space segment, ground segment, and user equipment development as needed to best deliver an interoperable, resilient, and synchronized total M-code solution into operations.[63]

---

[61] Resource loading a schedule is a program management best practice. A tutorial regarding the basic approach and benefits of resource loading a schedule can be found in Kim Koster, "What Your Schedule Isn't Telling You," paper presented at the PMI Global Congress 2011, Dallas, Tex., 2011.

[62] In a real-world example of the need for budgetary resilience, we originally intended to create a small example of the recommended sensitivity analysis and show how it could be used to understand the resilience of a project plan to budgetary uncertainty. However, our own budget was cut midyear. While we undoubtedly would have learned a great deal from creating such an example, it was a modular activity that could be removed from our project plan without excessive collateral damage to the larger research agenda. We note that this activity should be considered for future research.

[63] Dividing the GPS into separate Pes for space segment, ground segment, and user equipment is, in our view, a root cause for the lack of integration and synchronization of the program. The precision timing on which PNT accuracy is based is a tightly integrated interplay between the timing on the satellites, intersatellite links, space-to-ground links, and sophisticated algorithms in the ground segment that estimate and correct for bias and drift in each. Putting a funding barrier between those elements also put organizational barriers between them, decreasing the probability

Modularity Mitigates the Risk of Budget Instability

A well-known technique for improving system resilience to the inevitability and uncertainty of change is to modularize the system.[64] To say that a system is *modular* is a statement both about the degree of coupling (low) between parts and the degree of cohesion (high) within each part. A modular investment plan includes decoupled planning items that can be inserted or extracted without excessive collateral damage to the rest of the plan. Each modular item should also have high cohesion (i.e., all its activities or products should share a single purpose, such as the GPS M-code capability example described earlier). High cohesion will allow USSF to clearly communicate the capability that is added to or subtracted from the overall plan by each of the planning items.

Aggregation Mitigates the Risk of Budget Inflexibility

DoD and congressional budgeting processes are not designed for agility. As part of its PPBE process, DoD starts two years ahead of the actual budget submission to Congress with a review of national strategic priorities. Existing or planned program capability requirements are then vetted against the country's and DoD's strategic priorities to ensure that the requirements for those programs remain valid.[65] In response to valid requirements, USSF then begins formulating an input to DoD's budget request that is consistent with a top-line funding constraint for the service and prioritized on the basis of USSF strategic objectives tied to national strategies. After service approval, OSD makes final adjustments, as can the White House's Office of Management and Budget. Finally, the president's budget request is presented to Congress, where adjustments can be made by two committees in the Senate and two in the House of Representatives: the Senate and House Armed Services Committees (SASC and HASC) and the Senate and House Appropriations Committees for Defense. If the funding adjustments for any particular PE differ between the House and Senate versions, these differences are adjudicated during reconciliation. Only after the appropriations are signed into law, which does not always happen prior to the start of the fiscal year, are the funds available to USSF. Funding from Congress is made per PE and must be spent on that PE within a prescribed time frame, the

---

of successful technical integration. This partition does not need to be perpetuated; M-code and future enhancements to PNT capabilities could and in our view should be funded as an integrated PE.

In budgetary planning for new systems, we urge planners to consider alternate partitions of total system capability that will be more resilient to budget instability than the traditional partitioning among space, ground, and user equipment. In a distributed system, partitions of capability as a service are preferred to partitions based on the geographic location of the computers that host the capability.

[64] In fact, the use of modular systems to improve resilience dates back to ancient agricultures. Modularity "allows structurally or functionally distinct parts [of a system] to retain autonomy during a period of stress and allows for easier recovery from loss." National Park Service, "Modularity in Resilient Systems," webpage, last updated February 24, 2017.

[65] Even if Congress allocates funds to a program via the PE, those funds can only be spent in support of a validated requirement for that specific program.

duration of which is one to three years depending on the type of appropriation.[66] For USSF, facing unknown, ill-defined, or rapidly evolving threats, the probability that this long process will result in the right level of appropriated funds at the right time to be responsive to threats is not high.

One way to mitigate the risks related to the rigidity of the budgeting process is to aggregate related modular planning items into a larger PE, so that the reallocation of funds between the planning items is under USSF control. While Congress is reasonably skeptical of attempts to aggregate planning items into large PEs because it gives them coarser control of acquisition funding, we noted in our review of the USSF 2022 PEs that several small PEs that are functionally related and could be aggregated into more–moderately sized PEs.[67]

Understanding Budgetary Impacts on the Space Defense Industrial Base

USSF acquisition success depends on a healthy and innovative space defense industrial base. While the commercial sector that supports that base is currently thriving, it is notoriously cyclical and DoD will continue to have long-term monopsony power in the space economy for the foreseeable future.[68] While RAND research has extensively studied USSF synergies with the commercial space sector, for this report, we limit ourselves to the observation that if USSF is to be successful in leveraging the commercial space sector, it will need a systematized way to understand and communicate to Congress and DoD leadership the health of the overall space economy, the possible impacts of budgetary actions, and the investment needed to sustain the space economy.[69] Assured access to a healthy and innovative industrial base must be part of investment planning and Congress must be a part of that conversation.

To ensure consistent and accurate communication on this topic, USSF should develop tools, processes, and methods to assess (1) the resilience of USSF capability roadmaps to changes in the industrial base and (2) the impact of USSF acquisition strategies and/or budgeting decisions on the health of the overall market economy that sustains the base.[70] The quality of an

---

[66] Appropriations, also known as colors of money, include operations and maintenance (O&M); research, development, test, and evaluation (RDT&E); and procurement. Appropriations to these PEs must be spent within one, two, or three years, respectively.

[67] DAF, *Department of Defense Fiscal Year (FY) 2022 Budget Estimates: Air Force Justification Book Volume 1 of 1, Research Development, Test and Evaluation, Space Force*, May 2021.

[68] Monopsonies occur when there is only one buyer in a market. Because of the cyclical nature of the commercial space sector, the U.S. government is often the only long-term buyer in the market, giving it significant influence (Orkand Corporation, *Monopsony: A Fundamental Problem in Government Procurement,* Aerospace Research Center, 1973).

[69] Emmi Yonekura, Brian Dolan, Moon Kim, Krista Romita Grocholski, Raza Khan, and Yool Kim, *Commercial Space Capabilities and Market Overview: The Relationship Between Commercial Space Developments and the U.S. Department of Defense*, RAND Corporation, RR-A578-2, 2022.

[70] An example of such an analysis can be found in Bonnie L. Triezenberg, Colby P. Steiner, Grant Johnson, Jonathan Cham, Éder Sousa, Moon Kim, and Mary Kate Adgie, *Assessing the Impact of U.S. Air Force National*

acquisition strategy or budgetary action should be judged by assessing the full cycle of how acquisition decisions affect the market, and then how the entry and exit of companies in that market affect USSF's ability to deliver capability to the warfighter in a timely fashion.

Improving the Transparency of Congressional Interactions

More-transparent congressional interaction will be needed if USSF is to mitigate the impact and magnitude of budget instability. Figure 3.2 depicts our acquisition ecosystem map, highlighting interactions with Congress. In this figure, we have highlighted only the organizations that interact with Congress (shown on the far left) either directly or through OSD as an intermediary. The faintness of the connection lines indicates a lack of frequent interaction. The infrequency of these interactions, however, may not be an issue. A detailed study of Army acquisition funding that asked whether the volatility of program funding streams could be reduced by improved congressional relations found no correlation to frequency or type of congressional interaction.[71] The study found that even senior leader engagements did not produce less volatile outcomes, concluding instead that "consistent, honest, accurate, and timely" interactions are the key to successful congressional relations.[72] Of these four attributes, our recommendations focus on consistency, accuracy, and timeliness.

Security Space Launch Acquisition Decisions: An Independent Analysis of the Global Heavy Lift Launch Market, RAND Corporation, RR-4251-AF, 2020.

[71] Demarest, 2017.

[72] Demarest, 2017, p. 140. Note that both we and Demarest are defining *success* as reducing volatility in the difference between the president's budget request and the appropriation ultimately received from Congress for each PE.

Figure 3.2. United States Space Force Interaction with Congress

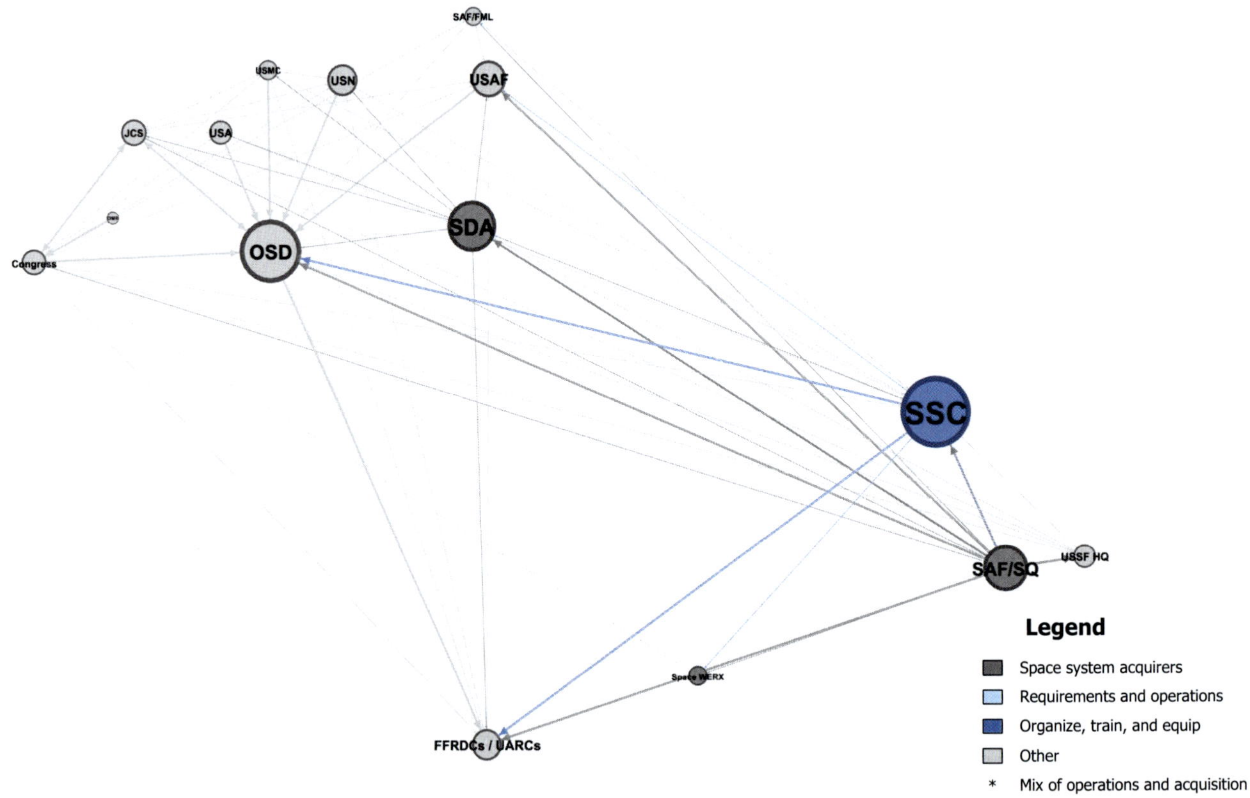

SOURCE: Features RAND analysis of interviews conducted in spring 2022, visualized using Gephi.
NOTE: Node and font size indicates the organization's centrality, calculated as the weighted sum of its interfaces. Line width indicates the strength of the relationship as recorded in our interviews. This diagram depicts organizational relations on the basis of interviews. Therefore, it reflects the as-is state of the ecosystem in spring 2022, prior to the SAE position being filled. In this figure, we have highlighted only the organizations that interact with Congress (shown on the far left) either directly or through OSD as an intermediary. The faintness of the connecting lines indicates a lack of frequent interaction.

Consistency speaks to the long-term nature of building trust but also to the need to develop a consistent understanding of acquisition strategies across USSF and the national space enterprise so that everyone tells and hears the same story. Accuracy is also a key element because when faced with inaccurate budget estimates, there is a tendency to suspect the accuracy of *all* estimates. Finally, timeliness is critical: At every step in the PPBE process, communication (or lack thereof) has fleeting opportunities to kill or advance the funding of an individual PE.

Our research found a lack of consistency and unity of effort both within USSF and across the U.S. space enterprise, which Congress has noted.[73] Congress is reasonably reluctant to allocate

[73] An example of that lack of consistency can be found in the capability planning for the missile warning and missile defense mission prior to the 2022 planning cycle. Originally, MDA, SDA, and SSC had competing plans, and NRO decided to go its own way to develop next generation overhead reconnaissance satellites. USSF then expended

funds without assurance of alignment across the space enterprise. To remedy this situation, USSF should (1) continue to improve USSF workforce education on budget processes involving the larger financial management community (e.g., DoD, White House, Congress) to allow them to more accurately assess and plan for the budget fluctuations and inflexibility of those processes, and (2) ensure that USSF personnel present USSF's strategy and plans for acquiring needed capabilities consistently to all external stakeholders, including Congress.

Longer term, USSF should implement a cultural change that strives for more open communication with Congress and builds alignment across USSF (the field commands and acquisition community), DAF, the Pentagon, and the greater U.S. space enterprise.

Reprogramming Authority

A short-term remedy to mitigate the inflexibility of the PPBE system may be to reprogram funds between PEs.[74] The current threshold at which funds can be reprogrammed between PEs without congressional approval is $10 million or 20 percent of the PE funding, whichever is lower. For perspective, $10 million is approximately 0.1 percent of the total fiscal year 2022 appropriation for USSF research, development, test, and evaluation.[75] In examining the fiscal year 2022 USSF PEs, we note that most PEs have funding levels below $200 million. For these PEs, $10 million or 20 percent of PE funding is probably an appropriate threshold. However, major acquisitions of space capability must be funded in the range of $500 million to $1 billion per year if that acquisition is to be completed in a reasonable time frame of three to five years.[76] For these major acquisitions, $10 million is unlikely to mitigate the types of issues that desynchronize the fielding of capability and a larger reprogramming of funding will be needed.

Although it is possible for USSF to request permission from Congress to reprogram funds from one PE to another if the threat has changed significantly or if conditions indicate allocated

---

significant effort to create an integrated MDA, SDA, and SSC program office to plan and manage efforts to modernize missile warning, missile defense capability. The architecture is now better harmonized with NRO planning. This is not to say that all inconsistencies have been resolved, but it is an indicator that USSF is moving in the right direction.

[74] As Eric Lofgren, research fellow at the Center for Government Contracting within the School of Business at George Mason University, has written:

> DoD needs a management structure that has modularity, iteration, speed, and competition built in from the start—not haphazardly provided for through reprogrammings. Portfolio management is the right long-term goal for agile acquisition. Increased reprogrammings can be a stepping stone to get there, but never the ultimate solution (Eric Lofgren, "Data: How DoD Reprogramming Has Changed (1961–2019)," Acquisition Talk, May 10, 2022).

[75] For the 2022 list of PEs and their funding levels, see DAF, 2021.

[76] This statement is based on an assumed total program cost of $1.5 billion to $5 billion. While many major space program acquisitions have had higher costs (for instance, GAO estimates that the next-generation overhead persistent infrared system will cost over $14 billion), we used this range to reflect USSF's stated desire to buy smaller systems (Jon Ludwigson, *Missile Warning Satellites: Comprehensive Cost and Schedule Information Would Enhance Congressional Oversight*, Government Accountability Office, GAO-21-105249, September 22, 2021b).

funding cannot be spent as originally planned, these reprogramming requests are rarely granted and the process of generating and approving the requests is quite slow and cumbersome.[77] Raising the reprogramming threshold, at least for the major programs, would allow the Secretary of the Air Force, with the advice of the USSF SAE and CSO, to retain access to the same overall level of funding while more quickly and easily directing that funding to the service's highest priorities in a timely manner. The likelihood that this mitigation is acceptable to Congress is positively correlated with improved congressional interactions.

[77] The low likelihood of approval has not always been true. In analyzing the results of reprogramming requests made from 1961 to 2019, Lofgren finds that in later years, approved reprogramming requests constitute less than 2 percent of the defense budget, while "between FY 1961 and FY 1971, DoD reprogrammed quite a bit more—between 2 percent and 8 percent of the DoD budget" (Lofgren, 2022). Not only is the process slow and cumbersome, the low likelihood of approval has a corollary because there is a high likelihood of losing funding for PEs identified as the source of funds, which disincentivizes requesters. Recent changes to allow requests without identifying the source of funding has reduced this disincentive.

Chapter 4. Architectural Focus

Many USSF organizations are involved in or with space-related architectural decisions; however, we found a significant lack of agreement on the roles and responsibilities of those organizations about who "owns" or has decisionmaking authority for elements of the space architecture, which is a significant challenge to achieving the integration and synchronization of those elements. Of particular concern, we found no organization with ownership of the capability architecture or roadmaps. While other U.S. military services use their PEOs for this function, the USSF PEOs are currently organized by acquisition organization rather than by capability.[78] Within SSC, the five PEOs *are* organized by capability but may not have sufficient insight into the innovation and experimentation taking place in other space acquisition organizations, hindering overall synchronization across experiments, rapid capabilities development, and PoRs. However, the USSF SAE has a small staff in the SAF/SQ office who could fulfill this function. Additional mission analysis, SoS architectural design and system engineering talent can be found within the SWAC and SSC's Space Systems Integration Office (SSIO).[79]

Table 4.1 summarizes our findings and recommendations regarding architectural focus: We will discuss each in depth in this chapter. The first recommendation recognizes that to formalize accountabilities for architecture, we first need to define *architecture*. Ideally, these definitions would then be used to document the roles and responsibilities of the various organizations involved in architecture in their mission statements and charters.

The second recommendation is to train the space acquisition workforce on the overall space architecture vision, definitions, organizational roles and responsibilities, and the relationships between organizations within the space architecture.

The final recommendation has short- and long-term components to ensure that the SAE has access to SoS architectural design talent when making acquisition decisions. In the short term, there are ways to assign architectural roles and responsibilities to ensure that the SAE has unfiltered access to SWAC and SSIO resources to support the small SAF/SQ office staff, but in the longer term, we recommend that USSF conduct a workforce study to determine how best to create and sustain organic mission analysis, SoS architectural design, and system engineering expertise within USSF. If USSF is to remain a lean and agile workforce, other organizational constructs might better serve their needs.

---

[78] There are five PEOs in SSC, one in SDA, and one in Space RCO. The five in SSC report to the USSF SAE, while the one in SDA reports to a board of directors, chaired by the Under Secretary of Defense for Research and Engineering, at least until October 1, 2022, and the Space RCO's PEO reports to the CSO.

[79] We do not mean to imply that additional USSF system engineering talent does not exist outside these organizations. There are many talented system engineers within SDA, Space RCO, and individual program offices. However, those engineers have a necessarily narrower focus than those in SAF/SQ, SWAC, and SSIO.

Table 4.1. Findings and Recommendations Related to Architectural Focus

Findings	Recommendations
There are different aspects of mission, system of systems, and systems architecture with responsibilities spread among multiple organizations.	Define what architecture consists of and use those definitions to document USSF mission statements and charters.
The space acquisition workforce does not have a consistent understanding of architecture or the roles and responsibilities related to architecture.	Educate the USSF workforce on architecture vision, definitions, responsible offices, and relationships among systems and missions within the architecture.
There is an insufficient amount of system-of-systems design talent within USSF, and yet the Service Acquisition Executive needs reliable access to that talent. It remains to be seen whether splitting the available talent between the Space Warfighting Analysis Center, the Office of the Assistant Secretary of the Air Force for Space Acquisition and Integration, and the Space Systems Integration Office is a sustainable strategy.	Short term: Clarify the Space Warfighting Analysis Center's and the Space Systems Integration Office's roles in defining system-of-systems architecture and advising the Service Acquisition Executive and the Office of the Assistant Secretary of the Air Force for Space Acquisition and Integration. While the Office of the Assistant Secretary of the Air Force for Space Acquisition and Integration may choose to outsource its need for system-of-systems engineering talent, advice should not be filtered through an external organization. Long term: Conduct a workforce study to determine how best to create and sustain organic system-of-systems and systems engineering expertise in USSF.

Defining Architecture

There are many definitions and uses of the term *architecture* across USSF. Our recommendation is that USSF develop a formal definition of the elements of architecture and promulgate it across the service.

For the purposes of this report, we partition the elements of space architecture into the following categories or views.[80]

- **Operational:** The operational view of architecture links military concepts to the employment of the technology used to execute military operations. It defines *how we fight* and is sufficient for developing acquisition strategies and identifying operational requirements that a solution must satisfy.[81] It does not specify an implementation.

[80] Architectures are commonly expressed using one or more *views*, and that is the term we will use here for the elements of an architecture. Note that architectural views are not different architectures. They are simply different views of a single architecture. For example, a building's architecture is often described with both an elevation view and a floorplan. Both describe the architecture of the building, just from different viewpoints.

[81] Note that this definition appears to be very close to what USSF calls *force design*, which is the stated responsibility of SWAC. We chose not to use that term in this report because we are concerned only with capability development activities and not with the full range of force design, as described in Joint Doctrine Note 2-19:

> Force design contains the Chairman's recommendations for innovation required to address mid-term to long-term challenges in the strategic environment. . . . force design spans both force

Operational views "describe the tasks and activities, operational elements, and resource flow exchanges required to conduct operations."[82]

- **Implementation:**[83] A capability is realized by acquiring a set of services, systems, or both. The implementation view describes the function of each service or system to be acquired and the relationships between them.[84] Ideally, the implementations map to the operational need and may have time-phased information to describe a planned evolution or technology insertions.
- **Capability:** The capability view describes the evolutionary path of interrelated implementations to "provide visualizations of the evolving capabilities so that Portfolio Managers can synchronize the introduction of capability increments across a portfolio of projects."[85] In this report, we call this description of the evolutionary path, containing both currently operational and funded and planned options, a *roadmap*. As described in the DoD Architecture Framework, the capabilities roadmap allows decisionmakers to "answer questions such as:
  - How does a particular capability or capabilities support the overall mission/vision?
  - What outcomes are expected to be achieved by a particular capability or set of capabilities?
  - What services [or systems] are required to support a capability?
  - What is the functional scope and organizational span of a capability or set of capabilities?
  - What is our current set of capabilities that we are managing as part of a portfolio?"[86]

Note that both the implementation and capability views answer the basic question, what do we fight with? However, the implementation view is designed to aid program management decisionmaking, while the capability view aids portfolio investment and synchronization decisionmaking.

- **Data and information:** The data and information views of the architecture answer the question: What C2 language do we use when we fight? These architectural views define data models from conceptual, logical, and physical viewpoints.[87]

---

development and capability development activities, with an emphasis on long-term solutions (Joint Doctrine Note 2-19, *Strategy*, Joint Chiefs of Staff, December 10, 2019).

[82] Chief Information Officer, DoD, "DoDAF Viewpoints and Models: Operational Viewpoint," webpage, undated-c.

[83] Another term that could be used for implementation is *solution*. We note that Space RCO favors the term solution when discussing the implementation of systems and services that meet operational need.

[84] As of July 2022, the DoD Architecture Framework description of services and systems viewpoints can be found at Chief Information Officer, DoD, "DoDAF Viewpoints and Models: Services Viewpoint," webpage, undated-d.

[85] Chief Information Officer, DoD, "DoDAF Viewpoints and Models: Capability Viewpoint," webpage, undated-b.

[86] Chief Information Officer, DoD, undated-b.

[87] The DoD Architecture Framework descriptions of the data and information viewpoints can be found at Chief Information Officer, DoD, "DoDAF Viewpoints and Models," webpage, undated-a.

Establishing Architectural Roles and Responsibilities

Per U.S. law, the Assistant Secretary of the Air Force for Space Acquisition and Integration shall "be responsible for and oversee all architecture and integration with respect to the acquisition of the space systems and programs of the armed forces."[88] Given that this position had not yet been filled by a Senate-confirmed appointee when the bulk of our interviews were conducted, it is perhaps not surprising that we found significant confusion regarding where the responsibilities for the various elements of the architecture lie. Table 4.2 summarizes the differing views we heard during more than 60 interviews with SMEs.

Table 4.2. Architecture Roles and Responsibilities Are Unclear to the Workforce

Architectural View	Who Thinks Who Owns What?
Operational: "How we fight"	• Space Systems Command believes the Space Warfighting Analysis Center owns operational architecture. • The Space Development Agency believes the Space Development Agency owns *certain parts* of operational architecture. • The Space Warfighting Analysis Center believes it owns operational architecture.
Capability: "What we fight with" roadmap	• The Space Warfighting Analysis Center believes it owns the *future* capabilities architecture, but not the current or the transitions. • The Space Warfighting Analysis Center believes the Program Integration Council owns the full roadmap. • U.S. Space Command believes the Program Integration Council *may be* responsible for the roadmap. • The Office of the Assistant Secretary of the Air Force for Space Acquisition and Integration believes the Space Warfighting Analysis Center owns the capabilities architecture. • Space Systems Command believes that it *has a role* in identifying future capabilities.
Implementation: "What we fight with" partition to systems and services	• The Space Warfighting Analysis Center believes Program Executive Officers may be responsible for determining the partitioning of capabilities to services and systems. • Program Executive Officers believe someone else is responsible; they simply execute the acquisition of systems and services assigned to them. • Space Operations Command believes Program Executive Officers are responsible, including for sustaining the implementation architecture. • Space Rapid Capabilities Office believes Space Systems Command is responsible for sustaining the implementation architecture.

[88] U.S. Code, Title 10, Section 9016, Assistant Secretaries of the Air Force.

Architectural View	Who Thinks Who Owns What?
Data and information: "The C2 language spoken in warfighting"	• U.S. Space Command believes the Space Warfighting Analysis Center is responsible for building the command control architecture. • Space Operations Command believes Space Systems Command is responsible for the command and control architecture. • U.S. Space Command believes the Program Integration Council is responsible for developing integration into the command and control architecture. • Space Systems Command believes it is responsible for common elements of the command and control architecture.

SOURCE: RAND analysis of more than 60 interviews conducted from January to May 2022.

In addition to defining the architectural viewpoints and who is responsible for producing them, the Assistant Secretary will need to define who is responsible for

- making decisions about the partitioning of capabilities and the systems and services that implement the capabilities
- measuring the progress of, variance from, or compliance with these architectural models
- proposing and implementing changes to the architecture to achieve the synchronized delivery and fielding of capabilities to operations.

Our findings related to these decisions, grouped by the architectural viewpoints that we defined in the prior section, are given in the following subsections.

Findings Regarding Operational Architecture Roles and Responsibilities

There is a fairly common understanding that SWAC owns the operational viewpoint for USSF architecture. However, we found significant concern among our interviewees regarding how the operational views are linked to requirements. More than one interviewee noted that the standup of the USSF SpOC as a field command and the creation of the Chief Strategy and Resourcing Officer (CSRO) had broken the process by which operational concerns are addressed during the requirements process. We will return to this issue in Chapter 7 in our discussion of the relationship between the acquisition and operational communities.

Findings Regarding Capability Roadmap Roles and Responsibilities

We found very little consensus on who is or should be responsible for creating and maintaining capability roadmaps. Because the DoD Architecture Framework clearly articulates that the capability views are specifically designed to assist in the portfolio management process, this is fundamentally a disagreement or lack of understanding regarding the responsibility for portfolio management. It is complicated by a lack of agreement over what comprises a capability and the fact that not all space acquisition organizations are contained within USSF. Possible solutions include

- allocating the task of *individual* capability portfolio management to the PEOs.[89] However, as we noted earlier, within USSF, only the SSC PEOs are organized by capability, and their portfolios are incomplete because they do not cover SDA or Space RCO systems and services that contribute to that capability.
- allocating the task of *USSF-wide* capability portfolio management to SAF/SQ. This would allow for complete portfolio management within USSF. However, even then, it is critical to acknowledge that for some mission areas, the portfolio may be incomplete because it does not include systems and services acquired elsewhere within DoD or the IC.
- allocating the task of *integrated enterprise* capability portfolio management to a Portfolio Acquisition Executive (PAE) who has acquisition authority for all space programs in DoD and the IC, as described in volume 3 of the *Report of the Advisory Panel on Streamlining and Codifying Acquisition Regulations* (also known as the Section 809 Report).[90] Appointment of a PAE is outside USSF's authority and therefore not a viable path forward in the near term.

USSF currently has two advisory entities—the SAC and the PIC—to mitigate the lack of an enterprise-wide portfolio manager; however, neither have acquisition authority. These advisory entities are described in the following subsections.

Space Acquisition Council

The SAC was established by Congress in the fiscal year 2020 NDAA to oversee, manage, and integrate DoD's space acquisition entities.[91] Its members include the Under Secretary of the Air Force, the Assistant Secretary of the Air Force for Space Acquisition and Integration (who serves as the chair and is also the SAE), the Assistant Secretary of Defense for Space Policy, the director of the NRO, the CSO, and the commander of USSPACECOM. The SAC is the formally recognized forum in which acquisitions decisions made by the SAE are certified. SAF/SQ, as part of its role in advising the SAE, could be charged with being the custodian of an enterprise-wide space capability roadmap.

---

[89] Note that to assign an individual capability to a PEO for portfolio management requires USSF to formally define its missions and capabilities or, as one interviewee expressed it, USSF needs to "decide where to put the seams."

[90] It is our understanding that the SAE's accountability is limited to DAF space systems and programs, although the position is also charged with promoting unity of effort with other national security space entities. Thus, the SAE is not a PAE as envisioned in the Section 809 Report. A detailed discussion of the panel's recommendation to transition from a program-centric (or even service-centric) acquisition model to a portfolio acquisition model can be found in Section 809 Panel, "Recommendation 36: Transition from a Program-Centric Execution Model to a Portfolio Execution Model," *Report of the Advisory Panel on Streamlining and Codifying Acquisition Regulations*, Vol. 3, January 2019, p. 53. We note that creation of a PAE as envisioned in the Section 809 report will necessitate a change in the acquisition chain of command of PM–PEO–milestone decision authority.

[91] The current charter of the SAC can be found in U.S. Code, Title 10, Section 9021, Space Acquisition Council.

Program Integration Council

The PIC has a rotating chair and is composed of leaders from NRO, MDA, the DAF Rapid Capabilities Office, Space RCO, SDA, SWAC, and SSC. The PIC does not have a currently approved charter and is, as of this writing, simply a forum for coming to agreement about which acquisition agency will supply which systems or services in a shared space capability. Actual acquisition authority resides with some of its members, but the agreements made at the PIC are not binding on those members. In Figure 4.1, we show the organizations that have acquisition authority in dark grey and their current relationships (if any) to the PIC. SSC's SSIO serves as the secretariat of the PIC and could be charged as the custodian of an integrated space capability roadmap.[92]

Figure 4.1. Program Integration Council Has Only Weak Interfaces Through Which to Coordinate Acquisition Activities

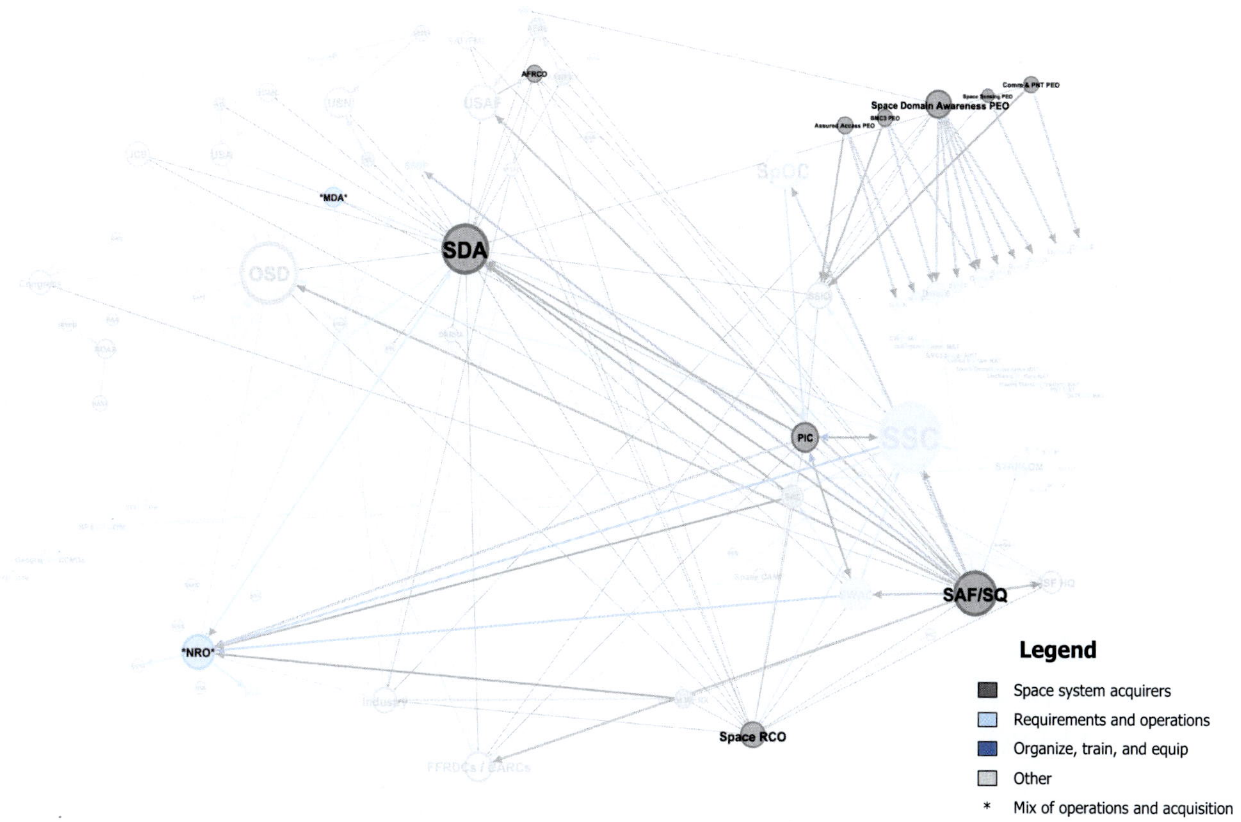

SOURCE: RAND analysis of interviews conducted in spring 2022 and visualized using Gephi.
NOTE: Node and font size indicates the organization's centrality, calculated as the weighted sum of its interfaces. Line width indicates the strength of the relationship as recorded in our interviews. This diagram depicts organizational relations on the basis of our interviews. Therefore, it reflects the as-is state of the ecosystem in spring 2022, prior to the SAE position being filled.

---

[92] The mission statement of the SSIO has not yet been approved as of July 2022. The envisioned role of the SSIO when initially stood up is described in Sandra Erwin, "Space Force Procurement Command Trying to Bring 'Unity of Effort' to Space Programs," *SpaceNews*, February 23, 2022a.

Findings Regarding Implementation of Architecture Roles and Responsibilities

The allocation of capabilities to systems and services is an output of the portfolio management process. PEOs today, in conjunction with their PMs, maintain implementation views of the systems that have been allocated to them to manage. Although the PMs clearly "own" these views of the architecture up until delivery of the system or service, it is much less clear who has the responsibility to sustain the implementation architecture following delivery. SSC has traditionally been the entity that contracts for the sustainment of delivered systems from contractors using operations and maintenance funding. SpOC, as the user of these systems to deliver capability, does not currently have the responsibility or funding to authorize enhancements or corrections to the delivered systems. While we found some cases in which SpOC commanders had taken the initiative to fund enhancements to their delivered systems, the lack of a current, accessible, and authoritative implementation view of the architecture that can be updated to reflect those enhancements is a cause for concern. This lack of a current, accessible, and authoritative view of the implementation architecture also complicates the transitions of rapidly developed systems from Space RCO to SSC for sustainment.[93] It is critical that USSF establish roles and responsibilities for the sustainment of implementation architecture views if it is to succeed in its quest to make systems and services more evolvable.

Findings Regarding Data and Information Architecture Roles and Responsibilities

Data and information should form the backbone of an integrated space architecture. However, we found little consensus regarding who is responsible for developing and delivering a data and information architectural view (i.e., data model). For the space domain awareness mission, SSC has developed a prototype of what is termed the Unified Data Library (UDL) to serve as a data broker, connecting different data sources to user applications. The SSC PEO for battle management and space domain awareness has stated that the UDL will evolve to serve as the backbone for the ATLAS system and will ultimately support the DoD's JADC2 architecture. It is possible that these efforts will yield an integrated space data and information architectural view. However, as of April 2022, the timeline for those efforts was reported as uncertain.[94]

Even if UDL becomes the backbone for space battle management and the protect and defend and space domain awareness missions, the more established missions would also benefit from a common data model, which we term the *capability data and information architecture view*. Consider, for instance, the satellite communications mission, which currently operates satellites acquired from different primes in different eras, each with their own proprietary C2 language. Currently, the 4th Space Operations Squadron operates the Defense Satellite Communications

---

[93] The issues associated with transitioning systems and services to operations are discussed in depth in Chapter 5.

[94] Frank Wolfe, "Timeline for Fielding Unified Data Library Uncertain, as U.S. Space Force Continues Prototyping," *DefenseDaily*, April 14, 2022.

System (Hughes Aircraft, 1970s era), Military Strategic and Tactical Relay (Lockheed Martin, 1990s era), Wideband Global SATCOM (satellite communications) (Boeing, 2000s era), and Advanced Extremely High Frequency (Lockheed Martin, 2000s era) satellite constellations, as well as the Enhanced Polar System (Northrup-Grumman, 2010s era). The lack of a common C2 language inhibits attempts to gain greater efficiency in operations by crosstraining operators on the different constellations. Given the high level of experimentation in this mission area (as noted in Chapter 2), developing a common data model for the satellite communications mission, even if started now, may be late to need. For the missile defense and missile warning mission, any standard that USSF develops as a C2 language must be harmonized with the MDA's C2 data model. Coordination with NOAA similarly will be required to develop data models for C2 of weather sensing satellites and payloads.

Recommendations Regarding Architectural Roles and Responsibilities

Given the confusion we found about how USSF manages architectural design, we recommend USSF define the roles and responsibilities outlined in Table 4.3 and document them in organizational mission directives and charters.

Table 4.3. Recommended Architectural Roles and Responsibilities

Role (recommended organization)	Responsibility	Approval Authority
Integrated operations architect (Space Warfighting Analysis Center)	• Create and maintain the integrated operations architecture views • Map operational needs to capabilities	Chief of space operations
Integrated space architect (to be determined—Space Acquisition and Integration direct report)	• Define the integrated set of capabilities to ensure complete coverage of USSF missions • Monitor the status of portfolio investment plans to – identify areas of overlap in USSF or the greater U.S. space enterprise – make recommendations to the Service Acquisition Executive regarding revisions to the definition of capabilities or the assignment of capabilities to portfolios to resolve issues	Service acquisition executive
Integrated data and C2 architect (to be determined)	• Create and maintain the integrated data architecture standards needed to ensure interoperability of USSF missions with one another, the joint force, and the greater U.S. space enterprise • Monitor the status of capability data and information architectures for areas of overlap or to identify gaps and resolve issues	Integrated space architect
Portfolio manager (Program Executive Officers)	• Create, maintain, and advocate for a resilient investment plan for the capability • Create and maintain the capability roadmap showing both current and future options for the capability • Maintain situational awareness of related non-USSF efforts and adjust the investment plan as warranted to maintain budgetary resilience and efficiency	Service acquisition executive

Role (recommended organization)	Responsibility	Approval Authority
Capability architect (deputy Program Executive Officers)	• Define the integrated set of systems and services that constitute the capability to ensure resilience and interoperability • Map Chief Strategy and Resourcing Officer–identified requirements to systems and services (i.e., implementations) • Monitor the status of system and/or service acquisition and sustainment activities to – identify gaps in the portfolio – identify systems or services that are no longer required – make recommendations to the portfolio manager regarding changes to the capability roadmap to ensure resilience and interoperability needs are met	Portfolio manager
Capability data and C2 architect (to be determined—within Program Executive Officers' organization)	• Create and maintain the data architecture standards needed to field an evolvable and interoperable capability	Capability architect
Implementation architect (Program Manager)	• Create and maintain the implementation architecture views for the assigned systems and/or services • Monitor the status of assigned systems and/or services acquisition and sustainment activities to – ensure Chief Strategy and Resourcing Officer–identified requirements are being fielded in a timely manner – ensure compliance with the data and information architecture	Capability architect

Assigning and Sustaining Architectural Talent

Architectural design talent in USSF is located in individual program offices: the SSC and SSIO, the SWAC, and SAF/SQ. In the short term, it should be possible to assign architectural roles and responsibilities to these offices without reorganization, so long as that assignment is clearly documented. For instance:

- SWAC could be formally chartered to create and maintain operational architecture views that correspond to the capabilities defined by the integrated space architect. We found very little dispute that it is currently filling this role and find that this is an appropriate assignment.[95]
- The integrated space architect role could be assigned to SSIO as part of its PIC secretariat role, SAF/SQ as part of its accountabilities to the SAE, or some still unidentified organization. A primary consideration is that the SAE has direct and unfiltered access to the integrated space architect.
- The integrated data and C2 architect role could be assigned to SSIO (where there are several efforts underway to fill that gap) or to staff in the Chief Technology and Innovation Officer's office. While the resulting architecture must be harmonized with

[95] The exception we found was at SDA, and we presume that the role of SDA versus SWAC will be resolved in the very near future, when SDA formally becomes part of USSF.

JADC2, we emphasize that the role as defined in this report is to create an integrated space domain (i.e., not all domain) data and information architectural view.

- As previously discussed, there are several candidates who could fill the portfolio manager role. If PEO portfolios are aligned with the capabilities defined by the integrated space architect (as opposed to aligned with the organizations that staff program offices), then having the PEOs take on the portfolio manager role appears to be the most straightforward option.
- The capability architect role, focused inwardly on the system engineering tasks of architecture and requirements allocation, should be kept separate from the portfolio manager role, which is focused outwardly on resilient investment strategies and advocacy. Furthermore, we are sensitive to SpOC's concerns about losing its input into the requirement development and allocation process. Given the frequent interactions we found between PEOs and the SpOC mission area teams, assigning the capability architect role to a deputy (or assistant) PEO and formalizing the capability architect's interactions with the SpOC mission area teams is a reasonable path forward. The capability architect's relationship to the CSRO also needs to be defined. A dual-hatted individual reporting to both the PEO and the CSRO is one solution. Figure 4.2 depicts the various elements the capability architect must harmonize to create the capability architecture.
- The capability data and C2 architect role could be assigned to SSIO or PEO staff (assuming PEOs are aligned with capabilities). It is essential that the role be filled for each of the capabilities: An integrated data and information architecture view for battle management is a necessary but insufficient part of what is needed to realize USSF goals for a resilient and interoperable space architecture.
- The role of implementation architect is currently filled by acquisition program offices. We agree this is the appropriate assignment of this role, but it must be expanded to include the sustainment phase of the implementation.

Figure 4.2. Context of the Capability Architecture

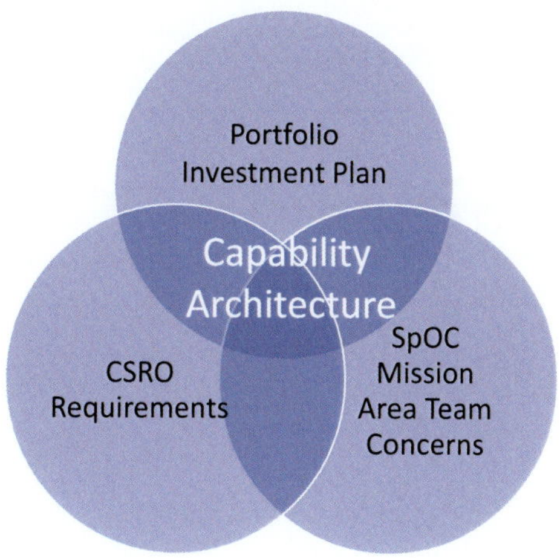

Longer term, splitting USSF's SoS architectural design talent between SWAC, SSIO, and SAF/SQ may not be a sustainable strategy. Not only is SoS system engineering talent a scarce resource, but it is also difficult to create an integrated SoS architecture without a thoughtfully structured design of the organizations that create that architecture.[96] Therefore, we recommend that USSF conduct a workforce study to determine how best to create, organize, and sustain an organic SoS and systems engineering expertise within USSF.

---

[96] The principle that technical architectures mimic the organizational structure of the entities that create them is known as Conway's Law: "Any organization that designs a system (defined broadly) will produce a design whose structure is a copy of the organization's communication structure" (Melvin E. Conway, "How Do Committees Invent?" *Datamation*, Vol. 14, No. 5, April 1968).

Chapter 5. Alignment of United States Space Force Organizations

Acquiring, fielding, and operating space systems requires coordination among many organizations within DoD, other U.S. government agencies, and international partners. Any failure to synchronize these organizations represents a clear risk in space acquisition. A shared understanding of roles, responsibilities, and authorities and clear lines of communication are vital to develop useful systems for the warfighter, make informed decisions on architecture design, and effectively synchronize fielding plans. Coordination among aligned organizations should help mitigate the major challenges facing USSF acquisitions, including responding to evolving missions, managing the acquisition cost of systems, building in cybersecurity, and planning for technology refresh and insertion.[97] It will be particularly critical in accommodating interoperability and in synchronizing organizations, incentives, and processes.[98]

The space acquisition environment is complex, diverse, and increasingly dynamic.[99] These attributes suggest that the space acquisition enterprise might benefit from decentralized decisionmaking (rather than centralized), coordination via collaboration (rather than standardization), and more organizational divisions (rather than fewer, more-integrated divisions).[100] However, these potential organizational design choices, some of which have been implemented or are underway, create new risks and may expose new problems if the stakeholders within the space enterprise are not coordinated.

---

[97] This list of major acquisition challenges is a subset of those identified in previous RAND work. See Jonathan P. Wong, Obaid Younossi, Christine Kistler LaCoste, Philip S. Anton, Alan J. Vick, Guy Weichenberg, and Thomas C. Whitmore, *Improving Defense Acquisition: Insights from Three Decades of RAND Research*, RAND Corporation, RR-A1670-1, 2022.

[98] *Interoperability* is defined here as the "ability to act together coherently, effectively, and efficiently to achieve tactical, operational, and strategic objectives" (DoD, *DoD Dictionary of Military and Associated Terms*, November 2021).

[99] Complex, diverse, and dynamic are three characteristics that should affect how an enterprise organizes, as described by organizational sociologist Henry Mintzberg in *The Structuring of Organizations* (Prentice Hall, 1979). The space acquisition environment is complex in that the problems faced by the space acquisition enterprise are often challenging and require specialized knowledge to solve, are diverse because of the wide variety of problems faced and activities conducted, and are increasingly dynamic in that the space acquisition enterprise is facing an accelerating pace of evolving problems.

[100] Mintzberg, 1979; Don Snyder, James D. Powers, Elizabeth Bodine-Baron, Bernard Fox, Lauren Kendrick, and Michael H. Powell, *Improving the Cybersecurity of U.S. Air Force Military Systems Throughout Their Life Cycles*, RAND Corporation, RR-1007-AF, 2015. We recognize this statement risks oversimplifying the organizational design challenge ahead of USSF and the wider space enterprise and that these characteristics must be balanced with other demands on USSF, including a desire for interoperability (perhaps through centralized decision authority) and the cost oversight required by senior DoD leaders and congressional appropriators. Nevertheless, these challenges have driven DoD and USSF to increasingly leverage nontraditional acquisition cultures and processes that rely on smaller, empowered acquisition groups.

Table 5.1 summarizes our findings and recommendations related to organizational alignment. We found that there was a general lack of understanding of organizational roles, responsibilities, and authorities across the enterprise. Much of the uncertainty stems from the rapid pace of organizational churn and can be lessened by formalizing mission statements and charters that clarify organizational intent—with a particular eye toward distinguishing the acquisition chain of command from OT&E responsibilities for the acquisition workforce. A lack of agreement on fundamental topics, including the partitioning of USSF missions, contributes to confusion on how organizations should integrate and synchronize. Harmonizing such functional concepts as much as possible would help increase understanding of how organizations align, even as personnel changes break and reforge working relationships. And because the operational community is often uncertain about how it can provide needed information to the acquisition community, making consistent options available through known processes and user agreements will help intentionally plan and clarify user engagement.

Table 5.1. Findings and Recommendations on the Alignment of United States Space Force Acquisition Organizations

Findings	Recommendations
Given the newness of USSF, many acquisition-related organizations lack defined and approved mission statements or charters. This leads to a general lack of understanding across the workforce regarding accountabilities, authorities, and touch points.	Formalize and approve mission statements to clarify and document • the acquisition chain of command versus organize, train, and equip responsibilities for the following organizations and suborganizations: Office of the Assistant Secretary of the Air Force for Space Acquisition and Integration, Program Executive Officers, Space Systems Integration Office, Space Development Agency, and Space Rapid Capabilities Office • the accountabilities and authorities (if any) for the Space Acquisition Council, Program Integration Council, and any warfighter integration councils • the roles of the Office of the Assistant Secretary of the Air Force for Space Acquisition and Integration mission area teams versus the Space Operations Command mission area teams • the role of Space Training and Readiness Command in (1) the generation of system requirements to support tactics, techniques, and procedures and (2) the developmental or operational testing of systems prior to fielding.
There is no standard partitioning of USSF missions or mission areas (i.e., portfolios of capability). Joint Publication 3-14, Office of the Assistant Secretary of the Air Force for Space Acquisition and Integration, Space Operations Command, and Space Systems Command all partition the missions differently.[a]	Harmonize the functional constructs to the extent reasonable and standardize mission area terminology.

Findings	Recommendations
While we found many venues for communications and feedback pathways between operators and acquirers, these are not clearly or consistently defined.	Continue to improve operator and acquirer communication processes: • Provide a known and consistent forum for operators to provide feedback to senior acquisition leaders (e.g., formalize the "pain point" process) • At the program level, draft a formal user agreement between user community and acquisition program office (akin to what is currently being used in the Adaptive Acquisition Framework software pathway).

[a] Joint Publication 3-14, *Space Operations*, Office of the Joint Chiefs, April 10, 2018, incorporating Change 1, October 26, 2020.

Defining Organizational Accountabilities and Authorities

USSF remains a new service that continues to determine its needs, strengths, and weaknesses and is making dramatic changes to address them all. Creating new organizations, or changing their scope and authorities, is an ongoing process that is part of the "most significant transformation in the history of the U.S. national security space program."[101] Chapter 2 discusses some of the major USSF organizations that have been created, and that discussion is not exhaustive. In addition to the new organizations, existing entities are undergoing changes to their internal organizational design, responsibilities, or authorities, and therefore how they interact with each other.[102] In addition, the communication of those changes to the enterprise is muddied by a legacy of how things have previously been done in Air Force Space Command and DAF headquarters. Put succinctly, things in the space enterprise are changing so quickly that many are struggling to keep up, and the number of relevant organizations makes it difficult and resource intensive to communicate with all parties involved.

In many cases, this difficulty is compounded by how the drafting and approving of mission statements and charters has often lagged the establishment or repurposing of an organization. While it is understandable when executing the mission is prioritized above ironing out organizational design, the continued absence of approved mission statements contributes to the confusion and lack of understanding that we found regarding organizational roles, responsibilities, and authorities across the enterprise. We found instances of entities with potentially overlapping or redundant responsibilities. In the following paragraphs, we highlight areas where more clarity is needed.

[101] DoD, *Defense Space Strategy Summary*, June 2020.

[102] For example, SMC became SMC 2.0 in 2018, then became SSC in 2021, and reorganized again in 2022.

Integration Councils

During our interviews, we often heard uncertainty or conflicting information about where and when touch points between different stakeholders were appropriate, needed, or helpful. In particular, the creation of multiple communication forums (such as the PIC and the Warfighting Integration Council) has led to confusion about the authorities (if any) of such forums. These forums facilitate communication but can blur the lines regarding who is empowered to make decisions, especially decisions that affect multiple programs.

Space Systems Command and Space Systems Integration Office

Unlike the air domain in DAF, the SMC commander was a PEO until SMC was reorganized to SSC in 2021.[103] The SSC commander now has OT&E responsibilities, along with a still-to-be-defined SoS integrator function.[104] This rather new transition from PEO to SoS integrator with OT&E responsibilities is creating ambiguity about chains of command and could affect decisive program execution by appointed PEOs and PMs.

SSC created the SSIO with broad but imprecisely defined responsibilities as a SoS integrator.[105] SoS do not integrate unless there is agreement from the PEOs and PMs who acquire those systems to use common standards, and our research shows that the SSIO has established relatively strong interactions with the PEOs (see Figure 5.1). However, it is unclear what authorities, if any, the SSIO has to make and enforce decisions or do anything other than perform in an advisory capacity and encourage PEOs and PMs to agree to use common standards. From this perspective and to more efficiently drive an SoS approach, the SSIO should be providing advice to the SAE, who is in the acquisition chain of command and can direct the necessary changes.[106]

---

[103] When Air Force PEOs were first established, they resided in Washington, D.C. Following a decision to move them closer to where programs were executed, the Air Force moved them to the field, and the commanders of the then–Aeronautical Systems Center, Electronic Systems Center, Air Armament Center, and SMC were PEOs. On October 1, 2001, SMC transferred to Air Force Space Command and the other centers remained in Air Force Materiel Command (AFMC) and were eventually subsumed into the Air Force Life Cycle Management Center in 2012 as part of an AFMC reorganization. During these changes in AFMC, PEO responsibilities were moved from the center commanders, thus separating the acquisition and OT&E roles. This was not the case for Air Force Space Command and SMC. The SMC commander remained a PEO until the 2021 SSC reorganization.

[104] As of August 2022, the appointment memo defining the roles and responsibilities of an SOS integrator had not been finalized and signed.

[105] Erwin, 2022a; Sandra Erwin, "Space Force to Reorganize Its Acquisition Command to 'Focus on the Threat,'" *SpaceNews*, March 4, 2022b; Jane Edwards, "Lt. Gen. Michael Guetlein on System of Systems Integration Within Space Systems Command," *ExecutiveGov*, March 7, 2022.

[106] Under the current construct, the SSIO advises the SSC commander who, in turn, advises the SAE. While this may work, it appears overly circuitous and potentially time consuming. From our understanding, it would be more efficient and potentially more effective if the SSIO reported directly to the SAE.

Figure 5.1. Connections Between the Office of the Assistant Secretary of the Air Force for Acquisition and Integration, Space Systems Command, Space Systems Integration Office, and Program Executive Officers

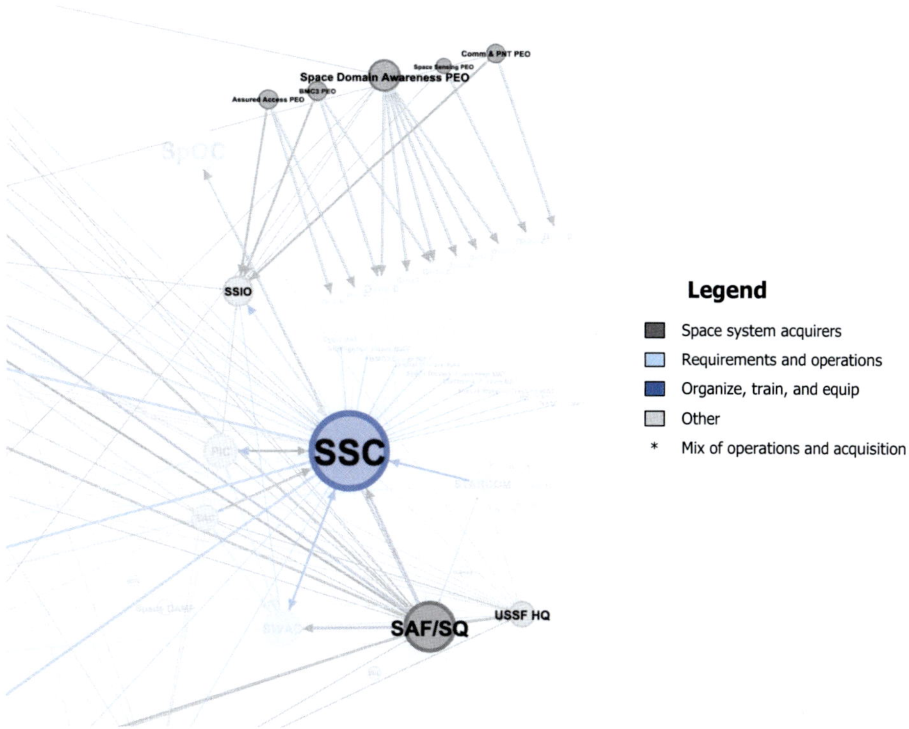

SOURCE: RAND analysis of interviews conducted in spring 2022 and visualized using Gephi.
NOTE: Node and font size indicates the organization's centrality, calculated as the weighted sum of its interfaces. Line width indicates the strength of the relationship as recorded in our interviews. This diagram depicts organizational relations on the basis of interviews. Therefore, it reflects the as-is state of the ecosystem in spring 2022, prior to the SAE position being filled.

Mission Area Teams

The role of Mission Area Teams (MATs), which currently exist in both the SAF/SQ and SpOC, should also be clarified. The SpOC and SAF/SQ MATs are partitioned differently (see the next section on "Harmonizing United States Space Force Missions"). The SpOC MATs appear, from our research, to be filling a critical role representing the operators' needs to the PEOs (see Chapter 7), and we can envision a future in which the SAF/SQ MATs fill a critical role providing capability or implementation architectures to the PEOs. Clarification of these roles and aligning them as much as possible will be helpful as both a part of intentional organizational design and a communication of roles and responsibilities to the space acquisition enterprise.

Space Training and Readiness Command

Many of those we interviewed reported that STARCOM's responsibilities in generating acquisition requirements are unclear. Part of this confusion is undoubtedly the nascent aspect of

STARCOM, even compared with other major parts of USSF.[107] STARCOM's central role in training, testing, exercises, and evaluations suggests that it should have some influence on related requirements. However, how this role will or can be leveraged by the larger USSF acquisition enterprise was not clear at the time of our research.

Need to Clarify Roles, Responsibilities, and Authorities

It is important that USSF move as quickly as possible to define, formalize, and publish the organizational responsibilities and authorities of organizations that participate in the acquisition enterprise. A clear delineation between the acquisition chain of command and OT&E accountabilities is particularly needed through mission statements and related guidance for the SAF/SQ, PEOs, SSIO, SDA, Space RCO, and their subordinate organizations. Organizational charters that describe the purposes and authorities of the various coordination forums (such as the SAC, PIC, and Warfighting Integration Council) will help clarify the roles of each of these entities.[108]

Harmonizing United States Space Force Missions and Mission Constructs

USSF has several missions affecting all domains of warfare that depend on projecting military power in and through outer space.[109] These missions are complex, multileveled, and often interdependent, making a standardized decomposition or partitioning of missions challenging to define.[110] However, a standard partitioning is necessary to support coordination among diverse stakeholders. Instead, we found that different function and mission constructs are used across the U.S. military space community. In some cases, several constructs exist within a single organization.

To determine how the stakeholders in the military space enterprise partition USSF missions, we reviewed the mission alignments of multiple organizations. Table 5.3 lists very common, somewhat common, and unique elements of mission partitioning constructs that are in use across the military space enterprise. These elements were identified and consolidated from 14 different sources spanning multiple space acquisition organizations, the operational community, and joint doctrine. Unsurprisingly, the mission constructs share some very common elements (e.g.,

---

[107] We conducted the bulk of our interviews within six months of STARCOM's inception, so some amount of confusion is to be expected.

[108] In the interest of not allowing the charter process to derail effective collaboration, we did hear advocacy for some forums be left unchartered. This may be particularly true for organizations (such as the PIC) where the goal is to coordinate outside USSF and across the larger U.S. government space enterprise. Within USSF, however, charters should be used to clarify the goals and purpose of each forum.

[109] For the purposes of this report, we partitioned USSF missions as described in Chapter 2.

[110] For example, theater and strategic missile warning are distinct missions *within* the overarching mission of missile warning, which is itself part of the missile defense mission. For example, the protect and defend space assets mission is supported by space domain awareness, which in turn is supported by space weather sensing.

satellite communications or PNT); however, some elements of the USSF mission set are only occasionally found (e.g., nuclear detonation detection or electromagnetic warfare) or are rarely discussed (e.g., fires or force protection). Even the "very common" elements are by no means standardized in the individual constructs that decompose USSF missions.[111]

Table 5.3. Frequency of Elements Referred to in Space Mission Partitioning Constructs

Very Common Elements	Somewhat Common Elements	Unique Elements[a]
Space domain and situational awareness	Space control (defensive and offensive)	Test and training (including evaluation and wargaming)
Position, navigation, and timing	Navigation warfare	Fire control
Satellite communications	Space-based sensing, environmental, and weather monitoring	Protection
Intelligence, reconnaissance, and surveillance	Nuclear detonation detection	Fires
Missile warning and tracking	Satellite operations	Net centric
Space list, access, and mobility	Cyber warfare	Force protection
Command and control and battle management	Electronic and electromagnetic warfare	Force application
	Orbital warfare	Force support

SOURCES: RAND analysis of multiple sources, including our interviews; Joint Publication 3-14, 2020; Space Development Agency, *Delivering Capabilities*, April 23, 2021; SWAC, *SWAC—Shaping the Force Design of U.S. Space Force*, December 17, 2021; Space Systems Command, 2021; SpOC, *Space Operations Command Mission Brief*, April 21, 2022.

[a] By unique element, we mean that only one of the mission partitioning constructs we reviewed called out an element by this name.

This inconsistency in how the USSF mission set is decomposed and defined leads to inefficient coordination, integration, and synchronization between parties in the military space enterprise. Organizational and functional seams are inevitable in a large and complex entity, but when those seams are placed differently by various organizations, implementing processes, or entities, managing those seams becomes increasingly difficult and potentially ineffective. Clear alignment between the PEOs and other stakeholders in the acquisition community (e.g., SWAC, SAF/SQ) versus the operational community to which they deliver capability (e.g., SpOC and STARCOM deltas and SpOC mission area teams) is particularly important. USSF should

[111] As part of our analysis, we tried to align similar concepts from policy, doctrine, and practice, but we were unable to do so. This indicated to us that USSF is currently undergoing a significant evolution in its thinking about the role of each mission, not simply an evolution in the vocabulary used to describe those missions.

harmonize functional and mission constructs to ensure that the right stakeholders are included in the appropriate mission-focused venues and forums.[112]

Providing a Known and Consistent Forum for Acquirer-Operator Interaction

Previous RAND research has long highlighted problems that can arise when the acquisition community is not well aligned with the operational community, especially when employing agile acquisition techniques.[113] The small size and technological dependence of USSF provides an opportunity to reduce the barriers between operators and acquirers.[114] Clear lines of communication between user and developer are critical to delivering usable and effective fielded capability. Prior research has also highlighted the importance of early engagement between users and developers, even during concept development, when such a relationship is not natural because the system is not in operations and sustainment.[115] Reimagining acquisition as a warfighting capability will be essential to evolving USSF to keep pace with advancing threats but will require a dramatic breaking down of the barriers between operators and acquirers.[116]

From our research, we found that better collaboration between operators and acquirers is inhibited by the following:

- There is uncertainty about how and when operators could provide input to the acquisition community or feedback to senior leaders. Issues related to organizational churn and the inconsistent decomposition of USSF missions feed this uncertainty.

---

[112] One such forum we learned of was a series of meetings between the SpOC MATs (organized by SpOC mission construct) and the SSC PEOs (organized by the SSC PEO mission construct) to discuss pain points. While we are highly supportive of such meetings, it is our observation there are significant ambiguities between the mission constructs used by SpOC and the SSC PEOs. The SpOC MATs are space domain awareness, space electronic warfare, missile warning, C2, cyberspace operations, ISR, SATCOM and navigation warfare, and orbital warfare. The SSC PEOs are assured access to space, battle management and C2, space domain awareness and combat power, communications and PNT, and space sensing. Given these different constructs, it may be difficult to determine who should attend which meeting. For instance, what SpOC refers to as cyber, electronic, and orbital warfare is highly entwined with what SSC refers to as battle management. One way to rationalize this ambiguity is to view battle management as primarily focused on centralized planning and SpOC's warfare constructs as the decentralized execution of that plan. However, this is not spelled out in any documentation that we had available to us.

[113] John Birkler, Mark V. Arena, Irv Blickstein, Jeffrey A. Drezner, Susan M. Gates, Meilinda Huang, Robert Murphy, Charles Nemfakos, and Susan K. Woodward, *From Marginal Adjustments to Meaningful Change: Rethinking Weapon System Acquisition*, RAND Corporation, MG-1020-OSD, 2010; Yool Kim, Guy Weichenberg, Frank Camm, Brian Dougherty, Thomas C. Whitmore, Nicholas Martin, and Badreddine Ahtchi, *Improving Acquisition to Support the Space Enterprise Vision*, RAND Corporation, RR-2626-AF, 2020; Wong et al., 2022; Mark A. Lorell, Julia F. Lowell, and Obaid Younossi, *Evolutionary Acquisition: Implementation Challenges for Defense Space Programs*, RAND Corporation, MG-431-AF, 2006; Jeffrey A. Drezner and Michael Simpson, *Exploring Parallel Development in the Context of Agile Acquisition: Analytical Support to the Air Superiority 2030 Enterprise Capability Collaboration Team*, RAND Corporation, RR-1808-AF, 2017.

[114] Shelton et al., 2021.

[115] Drezner and Simpson, 2017.

[116] Shelton et al., 2021.

- The PEOs and PMs are incentivized to fulfill their originally approved planning for cost, schedule, and performance; changing plans to accommodate new input from operators midway through a program is not necessarily rewarded.[117]
- There is a hesitancy on the part of some senior leaders to encourage their acquisition workforce to engage in career-broadening postings (such as having an operational tour) that might serve to break down some of the cultural barriers between the acquisition and operational communities.

Where we did find cases of successful integration between the communities, that success was driven by bottom-up personal relationships and a desire to use more-agile acquisition approaches for smaller efforts.

Providing a known and consistent forum for operators to provide feedback to senior leaders would reduce uncertainty about how and when significant operational concerns about the content or timing of fielded capabilities should be raised. For individual programs, expanding the use of formal user agreements (currently used for the software pathway) would improve lower-level integration between operators and acquirers and ensure that there is early engagement to define the role of operators in the systems and services being acquired. These coordination mechanisms may not be required for all programs (and other mechanisms may be used on an ad hoc basis as appropriate) but a consistent mechanism can be the cornerstone of better communication between operators and acquirers.

[117] In addition, Joint Requirements Oversight Council requirements ("big R" requirements) are often highly specific, inflexible, and form the primary basis for judging whether a program met performance goals—meeting so-called little r requirements direct from the user community is not similarly incentivized.

Chapter 6. Harnessing Space Innovation and Experimentation Through Transitioning to Programs of Record

In response to the growing threat to U.S. space systems and congressional demand for agility and efficiency in space acquisition, all USSF mission areas are experiencing considerable innovation through public and private sector funding.[118] Despite the movement toward increased innovation and experimentation, we found evidence that USSF is having difficulty harnessing this innovation. We identified some challenges and offer some potential solutions for USSF to harness innovation and experimentation and to transition this innovation and experimentation to PoRs. Within the space acquisition community, there is a need for transitioning innovation and experimentation from such organizations as Space RCO, SDA, SpaceWERX, and Space CAMP to SSC's PoRs.

Experimentation and Innovation in United States Space Force Acquisition Organizations

Harnessing innovation and experimentation requires synchronization and integration between requirements, budget and finance, acquisition, and operational communities.[119] Operators, PMs, and PEOs need visibility into current innovation and experimentation within the space community. USSF and stakeholders in space acquisition also need to be incentivized to accept the risks involved in integrating newer technology. An example of this would be the potential reluctance by PEOs and PMs to accept additional programmatic risk to existing programs by including newly demonstrated technology.

Experimentation and innovation are being carried out in several newer space acquisition organizations.[120] These organizations are organized differently from SSC—they are smaller and tend to be more focused on rapid, innovative solutions. They are as follows:

---

[118] A recent draft of the 2023 Senate NDAA again emphasizes this need and the concern that recent reforms, including the appointment of the USSF SAE, may be inadequate to meet these goals (Sandra Erwin, "Senate NDAA Directs Independent Review How the Space Force Acquires Technologies," *SpaceNews*, July 21, 2022f).

[119] The Oxford English Dictionary defines *experimentation* as "the action or process of experimenting or making experiments; a series of experiments" and an *experiment* as "the action of trying anything, or putting it to proof; a test, trial." The Oxford English Dictionary defines *innovation* as "the introduction of novelties; the alteration of what is established by the introduction of new elements or forms. . . . a change made in the nature or fashion of anything; something newly introduced; a novel practice, method, etc." While this chapter includes both, we will sometimes use innovation or innovate to refer to both for convenience. See Oxford English Dictionary, "experimentation, *n.*," webpage, undated; Oxford English Dictionary, "experiment, *n.*," webpage, undated; Oxford English Dictionary, "innovation, *n.*," webpage, undated.

[120] Note that this list is not exhaustive of all possible ways USSF could pursue innovative solutions.

- **Space RCO.** As described in Chapter 2, Space RCO was created by Congress in December 2017. The organization's mission, codified in law, is to focus on low-cost, rapid-reaction capabilities. Space RCO has announced that its first project, which was started in 2019, will be operational in 2022.[121] Space RCO has a dedicated PEO, a small size, and a limited number of programs. Given its urgent mission, the organization is staffed with highly experienced program management and functional support.
- **SDA.** Also described in Chapter 2, SDA was created by Congress in 2019 as a separate space agency within OSD. It is slated to transition to USSF in October 2022. SDA is currently committed to demonstrating capabilities in two-year increments, called tranches, with the first tranche due in 2022 and the second in 2024. Similar to Space RCO, SDA has a dedicated PEO and is able to be selective in its staffing.
- **Space CAMP.** Space CAMP is a software factory that works closely with warfighters to develop operational tools.[122] It has an atypical structure—it reports to the Air Force Research Laboratory (AFRL), it is funded by SpOC, and its staff positions are provided by SSC. The most widely known product of Space CAMP is the Platform One software development environment, which began as a Space CAMP initiative.[123]
- **SpaceWERX.** Part of AFWERX, a technology division of the AFRL, SpaceWERX connects operators with commercial, academic, and government organizations to develop innovative solutions and build the defense industrial base.[124] SpaceWERX's success is measured by whether the small businesses it supports go on to obtain contracts in support of USSF or other defense contracts, are acquired in mergers, or otherwise grow and invigorate the U.S. space industrial base.

Standing up organizations that focus on innovation and experimentation is only part of the solution for improving space capabilities. A persistent problem across DoD acquisition is transitioning technology from innovation and experiments to PoR and, ultimately, to the operational community (i.e., overcoming a proverbial valley of death). The literature is replete with discussion on this topic, including the following from the Defense Acquisition University:

> For many in the field of acquisition, there is a perhaps understandable idea that a great chasm exists between science and technology (S&T) and programs of record (PoRs) in advanced development. After all, this chasm traps many

---

121 Albon, 2022b.

122 Air Force Chief Software Officer, "Space CAMP," webpage, December 19, 2019.

123 Platform One enables flexible, responsive software development. For more information, see Platform One, homepage, undated. According to one interviewee, Platform One became very large and as a result of its success, had to be transitioned so that Space CAMP could continue with its primary focus of being responsive to operator requirements.

124 SpaceWERX, "About Us," webpage, undated. Interviewees noted that many people across DoD view getting new technology into a POR as the key measure of success in attacking the *valley of death*, a term used to describe the process by which a vendor or capability developer attempts to transition a product, prototype, or capability either to a DoD contract or into operational use. Efforts to build the industrial base and provide funding for small businesses to mature their technology enough for primes to see them as part of the supply chain are also metrics of success (Defense Acquisition University, "Transition Program," webpage, undated-d; James M. Landreth, "Through DoD's Valley of Death," Defense Acquisition University, February 1, 2022).

transitions, defined as the movement of data, knowledge, technology, and/or capability from S&T to advanced development.[125]

Throughout the literature and discussions conducted for this analysis, we documented a series of findings and recommendations regarding USSF's ability to better harness and transition innovation and the results of experiments into PoRs. These findings and recommendations are provided in Table 6.1, along with additional discussion on improving the status quo.

Table 6.1. Findings and Recommendations Related to Harnessing and Transitioning Innovation and Experimentation to Programs of Record

Findings	Recommendations
Significant experimentation is ongoing in multiple organizations devoted to innovation, but this experimentation is not visible centrally or across organizations.	Require organizations involved in space acquisition to improve visibility into their efforts and ensure capability roadmaps include all experiments and innovations.
• Acquirers are incentivized at the individual program level instead of the enterprise level, which limits willingness to incorporate new or emerging technology.	Implement a cultural change that values enterprise and mission success over personal or individual program success, including • aligning Program Executive Officer and Program Manager incentives with desired outcomes • developing more-flexible and mission- or enterprise-focused program requirements—for example, using operator pain points or broad thrust areas.
• Innovations are typically not accompanied by incorporation funding.	• Develop criteria for when to terminate aging systems and delayed programs and require justifications for exceptions, i.e., make it okay to terminate programs. • Reallocate funding from terminated or underperforming activities to higher priorities, i.e., make it okay to give up funding.
There may be insufficient acquisition personnel with the skills and experience to readily adapt innovation and experimentation into programs of record.	Investigate options for recruiting and retaining acquisition talent.

Overcoming Challenges in Harnessing Experimentation and Innovation

We found evidence of several challenges to harnessing and transitioning innovation successfully to PoRs. These challenges fell into four broad categories: visibility, culture, process, and personnel. Our interviews provided some specific examples of where innovation was not being transitioned into programs or operations, including the following:

- One product developed by an innovation organization was well liked by the operator, but the materiel leader rejected pursuing the solution because another organization was under contract to develop something similar. While this may have been because the item under

[125] Richard Newton, "Bridging the Valley of Death and Thriving Beyond—Improving Transition Success with Metrics," Defense Acquisition University, July 30, 2022.

contract was being designed to be more sustainable or interoperable than the existing innovation, our interviewee was unaware of such concerns.

- Another innovation organization transitioned an application to a program office, which unsuccessfully tried to transition the application to its own platform. This lack of success may have been because the integration complexity was underestimated or there was inadequate planning for interoperability.
- One innovation organization noted that it was starting to see innovations flounder when moved to larger organizations. Another said it was very difficult to find a program office with the "right fit" to successfully adopt and sustain innovative applications. In these cases, the implication was that finding the right fit was primarily a cultural issue.[126]

Increasing Visibility into Space Innovation

A significant challenge to harnessing innovation and experimentation is the lack of visibility into the many ongoing efforts across USSF and, in some cases, across the space enterprise.[127] The sheer number of government and private organizations innovating in the space domain makes it difficult to maintain visibility. To address this, we recommend better communication throughout USSF. This may include the SAE requesting all USSF organizations involved in space innovation to improve visibility into their efforts. Given the many organizations outside USSF involved in space-related innovation, communication mechanisms should extend beyond USSF to include other federal government agencies. As we recommend in Chapter 4, to achieve greater visibility, capability roadmaps should include not just PoRs but also known experiments and innovations. These capability roadmaps should include optional off-ramps or on-ramps that are based on a variety of possible outcomes from the experiments and innovations.

With respect to better engagement with industry, USSF has created many forums that have resulted in some confusion. While such initiatives as SSC's "Front Door," designed to provide industry with a single interface to all USSF innovation forums and activities, are promising, as of September 2022, it is too early to assess the extent to which such efforts are yielding results.[128] Fewer and more-focused forums are recommended.

Finally, we note that although small or nontraditional companies are supported through Small Business Innovation Research (SBIR) programs and other research and development authorities, they are often reliant on close relationships with prime contractors to effectively transition their component or subcomponent capabilities for integration into larger systems.[129] We learned of the

---

[126] Please see Shara Williams, Jeffrey A. Drezner, Megan McKernan, Douglas Shontz, and Jerry M. Sollinger, *Rapid Acquisition of Army Command and Control Systems*, RAND Corporation, RR-274-A, 2014, for additional background information.

[127] For example, with the increased availability of very small satellites to conduct experiments, the number of organizations conducting such experiments has increased. Also, consider AFRL's experimentation with antijamming techniques that we describe in Chapter 1.

[128] See, for instance, Theresa Hitchens, "Companies Knock, but Space Force Acquisition Portal 'Front Door' Remains Closed," *Breaking Defense*, July 7, 2022b.

[129] Shelton et al., 2021.

use of SBIR programs as a parallel path to support the development and transitioning of critical technologies into larger prime contracts, and we encourage USSF to do more of this.

Changing Culture by Aligning Incentives

Failure to properly recognize and mitigate the significant risks represented by a lack of resilient, interoperable, and rapidly evolvable space systems threatens national security. While this threat is recognized by Congress and by USSF leadership, program and personnel evaluation systems continue to incentivize individual goals above enterprise goals. An unwillingness to take on additional cost, schedule, or performance risk in PoRs is cited as one reason why innovation does not get harnessed. According to the Aerospace Corporation, this is the "the high-reliability, low-risk mindset that discourages change, avoids risk-taking, and stifles innovation."[130] Prior RAND analysis has also described risk aversion in DoD acquisition organizations.[131]

Our recommendations for addressing these cultural challenges are to encourage intelligent risk taking and align incentives to desired behaviors. We note that USSF appears to be improving on the first of these recommendations but has made little noticeable progress on the second. Several interviewees noted that an agile service will need to be better at balancing risk and that there can be good efforts and good risks that still result in failure. However, the existing culture needs to change to allow PMs and materiel leaders to "[take] smart risks" and not punish failure, should that be the result. In other words, acquisition professionals will need to believe that taking risks that sometimes fail—perhaps because of changes in the threat or in warfighter needs—will not reflect on them personally or harm their careers.

If USSF wants to more effectively harness innovation and transition that innovation to PoRs, leadership should implement a cultural change that values enterprise and mission success over personal or individual program success. Government programs and thus PMs are evaluated individually and incentives focus on program cost, schedule, and performance. This means that harnessing innovation, implementing enterprise thinking and performance, responding to operator needs, and delivering capabilities that address the threat are not as strongly or universally incentivized. In fact, we were told by interviewees that there is little incentive for PMs to look beyond cost, schedule, and performance, and USSF does not always reward doing what is best for the entire enterprise. Two related threads ran through this discussion. One was to encourage a broader focus—centered on enterprise performance and mission success—rather than individual program success. The other was to decouple individual success from program success. Suggested changes include:

---

[130] Project Thor Team, *Outpacing the Threat with an Agile Defense Space Enterprise*, Aerospace Corporation, September 2019, p. 4.

[131] See, for example, Shelton et al., 2021, pp. 11, 17. Risk aversion can also slow the pace of the acquisition process, which is another challenge described later in this chapter. See also Wong et al., 2022, p. 34.

- developing alternative criteria for assessing program success, such as delivery and user acceptance or whether a program delivered an operationally relevant system that addressed the current threat
- providing more coaching and context for all personnel to understand the big picture
- incentivizing a perspective and actions that focus on enterprise performance, such as interoperability, total SoS resilience, or SoS evolvability
- incentivizing delivery of capabilities instead of programs and systems.[132]

From our research, we conclude that the USSF workforce's perceptions regarding which behaviors are incentivized have yet to change in any significant way. One way to more concretely indicate to all personnel that USSF is serious about incentivizing different behaviors in PEOs and PMs would be to update the current Air Force instruction that describes the integrated life cycle management objectives, roles, responsibilities, and evaluation criteria.[133]

Changing Culture to Meet the Requirements and Funding Process Challenges

Recent analyses have noted the importance of a responsive acquisition system that provides capabilities quickly to meet evolving threats.[134] Interviewees reiterated this challenge on multiple occasions. As noted in our discussion in Chapter 3 on budget resilience, much constrained agility can be traced to the PPBE process. However, overly specific and enduring requirements and an overall lack of funding also affect USSF's ability to achieve budget resilience. These are discussed in greater detail in the following subsections.

Requirement Inflexibility

Overly specific and difficult-to-change requirements pose another barrier to harnessing innovation.[135] Standard practice is to write system requirements that are "SMART"—specific, measurable, attainable, realistic, and timely.[136] But keeping requirements SMART in a rapidly changing environment can be achieved only with a flexible but rigorous change process.

---

[132] Interviewees noted that a program could meet its schedule but might still not address the threat because of the speed at which a threat can evolve and the length of time the acquisition system can take.

[133] Air Force Instruction 63-101/20-101, *Integrated Life Cycle Management*, DAF, June 30, 2020, incorporating Change 1, November 23, 2021, governs USSF acquisition activities. Interoperability with allied systems is an evaluation criterion and is briefly mentioned as a criterion for U.S. SoS, but there is no mention of resilience or evolvability. The term *interoperability* appears 29 times, but many of those instances are about modeling and simulation, not operational systems. In contrast, the terms *cost* or *budget* appear more than 300 times and *schedule* appears 100 times. If DAF cannot update the Air Force instruction in a timely fashion, USSF, as a separate service, could issue its own supplemental instruction.

[134] Shelton et al., 2021, p. 15; Ludwigson, 2021a, p. 2; Project Thor Team, 2019, p. 7.

[135] Dissatisfaction with the lack of flexibility in the DoD requirements development process is not unique to USSF.

[136] An example of a SMART requirement is "The system availability under contingency conditions shall be greater than 99.5% over any one-year period." Delineation of the contingency conditions would improve the requirement further. An early paper on SMART can be found by George T. Doran, "There's a S.M.A.R.T. Way to Write Management's Goals and Objectives," *Management Review*, Vol. 70, No. 11, November 1981.

Additionally, flexibility and rigor in processes are often at odds and requirements often need to be balanced against one another. The contrast of security and usability of systems is a classic example of opposing requirements: The most secure system is one that is unusable and security controls often impede systems usability.

The balance between requirements also changes in response to the threat. Developers of rapid capabilities may make short-term trades that a developer of a more enduring capability would not. Too specific or too narrowly defined requirements may allow the acquisition community to jump to the point solution rather than take the time to understand the trades. Existing requirements can handcuff innovation or prevent trade-offs that allow innovative implementations that might better satisfy warfighter needs (e.g., an innovative sensor technology emerges with significantly better accuracy but with a slightly lower availability than the requirement). Specifying performance requirements as an envelope with edges that are defined by a range of acceptable values is one way to maintain specificity while allowing flexibility in implementations and increasing trade space.[137]

Some interviewees described a different approach to writing requirements that is being used by innovative organizations (e.g., SpaceWERX) and the operational communities they support.[138] Rather than writing specifics about what the system shall do or the thresholds it should meet, requirements are instead expressed in terms of pain points that operators experience while doing their jobs.[139] Any solution that alleviates a pain point is acceptable. For areas in which USSF has a strong need for innovative solutions, writing requirements as pain points to be alleviated or as an envelope of allowed performance values could improve the harnessing of innovation in two ways. First, it would allow the business community to be creative rather than driving it to a particular solution. Second, defining an envelope of performance requirements enables trade-offs.

Funding Challenges

While USSF has stood up several labs to pursue innovation, available funding to transition that innovation to programs is lacking. Interviewees noted that PEOs do not have adequate funding to transition prototypes and new technology from the labs. USSF also struggles with achieving a balance between older, legacy systems that need continued PEO and financial support and newer technologies and capabilities that will be integrated into a larger system or on-

---

[137] Specifying performance as the edges of an envelope is a common practice in the aircraft and building industries. It is less common in the space industry.

[138] SpaceWERX has helped small businesses be more responsive to operators' needs by proposing innovative technology solutions (authors' interview with USSF official, March 31, 2022). For additional information, see AFRL Public Affairs, "SpaceWERX Ready to Propel Space Innovation," Air Force Research Laboratory, August 25, 2021.

[139] A quick read on how analysis of pain points leads to improved systems can be found at "Analysing Business Process Pain Points to Get Better Requirements," Aotea Studios, 2011.

ramped as part of the program. Below, we explore some of the funding challenges that contribute to this issue.

Inadequate funding of innovation: During a hearing of the House Armed Services Subcommittee on Cyber, Innovative Technologies, and Information in May 2022, a congressional member noted that "innovation is chronically underfunded in the DoD budget."[140] For example, the fiscal year 2023 DoD budget request for science and technology is 13 percent lower than what Congress authorized for fiscal year 2022 ($18.8 billion to $16.5 billion), and the budget request for basic research decreased by 14 percent ($2.6 billion to $2.4 billion).[141] One interviewee observed that there is not enough science and technology funding in the budget and that innovations exist that could be handed off for use, but such organizations as SSC do not have the funds to pick up new technologies.

Inability to cancel programs: Another reason for the lack of available funding is the difficulty canceling programs that are no longer needed. Interviewees described programs that are inadequately funded (or where funding has been delayed for many years), indicating a lack of demand for the capability, yet no one will actually cancel the effort.[142] They noted that USSF cannot afford to buy and sustain everything that is needed, so USSF leadership needs to make tough decisions about committing to fully fund some systems while canceling others. The reasons cited for not canceling programs include: (1) prime contractors will contact Congress to challenge cancellations, even when the operational environment has changed; (2) program cancellations are perceived as failures instead of money savers; and (3) decisionmakers are unable to objectively assess whether a funded program provides needed benefits and/or whether the technology is mature enough to continue. As we noted in Chapter 3 when discussing budgetary resilience, objective, evidence-based studies regarding DoD budgetary impacts on the health of the defense industrial base can provide a much needed perspective on the impact of cancelling programs. USSF should also develop objective criteria for terminating aging systems or programs that have repeated delays, assessing each program against those criteria and terminating any that meet those criteria. If a program that meets the criteria is not terminated, decisionmakers should provide a justification for retaining it.[143] Providing objective criteria that are consistently applied would improve transparency and decisionmaking and would be useful

---

[140] Sean Carberry, "Congress Alarmed over Innovation Funding," *National Defense Magazine*, July 6, 2022.

[141] Carberry, 2022.

[142] The terrestrial weather sensing mission appears to fit this description. Despite the extensive overlap between military and civil earth weather sensing, DMSP satellites are kept operational well after their mission design life is over and, in some cases, after multiple onboard failures (the remaining four satellites were launched between 1999 and 2009 and had a five-year design life). DMSP replacements have been proposed several times since the mid-1990s but have never come to fruition. Yet, USSF is reluctant to give the up the mission (William McCormick, "DoD Plans to Replace DMSP Weather Satellites Within Five Years," *ExecutiveGov*, May 24, 2021).

[143] This is like the existing Nunn–McCurdy process. The Secretary of Defense must write a memo to Congress documenting justification for not cancelling a program that meets the objective Nunn–McCurdy thresholds.

when industry or Congress questions decisions. Criteria could include age of the technology, whether other systems provide the same or better capabilities, how well the technology addresses current or expected future threats, and warfighter perceptions of usefulness.

Inflexible funding: Even when funding is available, getting it to where it is needed quickly can be challenging.[144] Two related issues, both discussed at in Chapter 3, constrain funding agility:

- DoD's PPBE process is not designed for agility. As others have noted, funding

  > a sudden emerging need or new technology maturation within the next two to three years would require intervention at the highest levels of the DoD, possibly through a reprogramming decision for the current budget or a resource management decision in the next year's budget.[145]

- There are constraints in shifting existing funding. Several interviewees mentioned the lack of funding flexibility, a sentiment reiterated by Air Force General Arnold W. Bunch, Jr., in his recent exit interview with *Air and Space Forces Magazine*.[146] To take advantage of a recently emerged innovation or successful experiment, current practice makes it likely that USSF would need to move funds between Pes.[147] However, congressional approval is needed to move more than $10 million dollars or 20 percent of PE funding (whichever is less). For perspective, $10 million dollars in 2022 is roughly equivalent to a 20-person engineering effort for one year (a relatively low threshold), and reprogramming requests can take months to adjudicate because actions happen only twice a year. Continuing resolutions can cause even further delays.[148]

Outside the formal reprogramming process, PEOs and PMs can consider several potential alternatives. The simplest approach may be for PEOs and PMs to reserve funds for onboarding new technology. However, management reserve funds are often consumed by other issues, and it becomes difficult to reserve funds if the technology is not truly ready. As an alternative to program reserve, as we noted in Chapter 3, creating Pes that cover broader portions of a mission area or capability portfolio would allow portfolio managers to shift funding to adapt to changing threats and to take advantage of new innovations.

---

[144] Wong et al., 2022, pp. 6, 27; Shelton et al., 2021, p. 19; Project Thor Team, 2019, p. 7.

[145] William Greenwalt and Dan Patt, *Competing in Time: Ensuring Capability Advantage and Mission Success Through Adaptable Resource Allocation*, Hudson Institute, February 2021, p. 43.

[146] Arnold W. Bunch, Jr., quoted in John A. Tirpak, "Bunch Says Quick-Reaction Funds Would Help New Tech Cross the 'Valley of Death,'" *Air and Space Forces Magazine*, June 10, 2022.

[147] The defense budget is made up of thousands of PEs and funding is assigned to each PE. For more information on the defense budget structure, see Brendan W. McGarry and Heidi M. Peters, "Defense Primer: Future Years Defense Program (FYDP)," Congressional Research Service, IF10831, version 10, December 15, 2021.

[148] For additional information on reprogramming, see Brendan W. McGarry, "Defense Primer: DoD Transfer and Reprogramming Authorities," Congressional Research Service, IF11243, December 21, version 6, 2021.

Sustaining the Space Acquisition Workforce

Concerns about whether DoD has enough acquisition staff with sufficient skills and experience predates USSF by many years.[149] In 2011, GAO noted concerns about the numbers of space system acquisition personnel and their experience levels.[150] A 2019 GAO report stated that

> [i]t is unclear whether DoD has a sufficient workforce to manage multiple new space programs. We issued a report last month that found DoD did not routinely monitor the size, mix, and location of its space acquisition workforce. . . . Echoing concerns raised in our prior work, we also found that DoD had difficulty attracting and retaining candidates with the requisite technical expertise. Officials from the Air Force's Space and Missile Systems Center were concerned that there are not enough experienced mid-level acquisition personnel and also expressed concern that the bulk of military personnel assigned to program management positions were more junior in rank than the Center was authorized to obtain.[151]

USSF interviewees confirmed these concerns, citing insufficient numbers of contracting staff with the right competencies and experience, the inability to retain contracting staff in the face of competition from organizations that offer full-time telework, competition for engineering staff from private industry and support contractors that can pay more, competition from other military and government organizations that can offer higher-level positions, and the government's slow hiring process.

Interviewees from innovation organizations frequently stated that their organizations' success depends on having competent and experienced staff in both technical and functional areas. Extending those models to larger organizations to foster innovation could be a tall order, given that many already have trouble recruiting and retaining such staff. USSF could consider the following options to help address workforce availability and competency:

- evaluate positions to broaden telework where possible
- investigate recruiting and retention bonuses for civilian (and military) personnel in high-demand specialties (e.g., artificial intelligence, contracting, engineering)
- reassess positions to determine where grade increases are justified
- investigate the benefits and feasibility of training programs (e.g., paying for education in exchange for service commitments)
- consider a technical track for officers that will allow promotion without command.

While the incentives and options for alleviating workforce challenges are not new suggestions, a greater focus on pursuing these options is necessary, given the importance of

---

[149] See, for example, Wong et al., 2022, pp. 21–22.

[150] Cristina T. Chaplain, *Space Acquisitions: DoD Delivering New Generations of Satellites, but Space System Acquisition Challenges Remain*, testimony before the Subcommittee on Strategic Forces, Committee on Armed Services, U.S. Senate, Government Accountability Office, GAO-11-590T, May 11, 2011, p. 27.

[151] Christina T. Chaplain, *Space Acquisitions: DoD Faces Significant Challenges as It Seeks to Address Threats and Accelerate Space Programs*, Government Accountability Office, GAO-19-482T, March 27, 2019a.

harnessing and transitioning innovation and experimentation into new space capabilities to keep pace with the threat.[152]

[152] Additional insight into the challenges faced by the space acquisition workforce can be found in Lauren A. Mayer, Mark V. Arena, Frank Camm, Jonathan P. Wong, Gabriel Lesnick, Sarah Lovell, Edward Fernandez, Phillip Carter, and Gordon T. Lee, *Prototyping Using Other Transactions: Case Studies for the Acquisition Community*, RAND Corporation, RR-4417-AF, 2020.

Chapter 7. Transitioning Technology from the Space Acquisition Community to the Operational Community

Our research revealed two valleys of death associated with fielding synchronized USSF capabilities to the warfighter. The first, transitioning the results of experiments and rapid capability development efforts to PoRs, was discussed in Chapter 6. The second, transitioning developed capabilities from the acquisition community to the operational community, is discussed in this chapter. The latter specifically refers to effectively fielding capabilities from Space RCO, SDA, SSC, and other organizations to the space operational community. This requires synchronizing and integrating newer capabilities with legacy systems that are already operational, which involves planning throughout the acquisition life cycle for the eventual transition.

USSF PEOs, the development community (e.g., SDA and Space RCO), and SpOC MATs, along with the functional communities supporting the acquisition community (e.g., testing, cybersecurity, financial management, contracting), play important roles in preparing for technology transition. Although we found significant improvement compared with prior years regarding integration between these entities, we also found persistent issues with transitioning technology into operations.[153] Table 7.1 summarizes our findings and recommendations related to the transitioning of capabilities into operations. It should be noted that all of these findings and recommendations are also discussed in prior chapters, but here we provide additional perspective on their impact on the ultimate goal of fielding synchronized capability into operations.

Table 7.1. Findings and Recommendations Related to Transitioning Technology from the Acquisition Community to the Operational Community

Findings	Recommendations
A roadmap identifying opportunities to insert technology into operations does not exist.	In addition to creating the capability roadmaps discussed in Chapter 4, create a clear shared vision between acquisition and user communities as to how and when technologies should be integrated into the broader capability architecture.
Resources needed to transition capability into operations are underallocated and current incentives can lead to the development of obsolete capabilities or capabilities that are never transitioned.	As discussed in Chapter 6: • Preserve funding and timing to on-ramp or rapidly transition technology that emerges during acquisition into ongoing development projects and/or operations • Establish a process to stop funding capabilities and technologies that fail to meet operational needs

[153] See, for instance, Shelton et al., 2021.

Findings	Recommendations
As noted in Chapter 5, while we found many venues for communications and feedback pathways between operators and acquirers, they are not clearly or consistently defined.	Develop and formalize a process to incorporate warfighter feedback into acquisition decisionmaking, as described in Chapter 5. This includes: • establishing a lead entity and/or organization to orchestrate the process • establishing a semiannual or quarterly battle rhythm event for warfighters and developers to discuss ideas, pain points, and requirements • establishing a clear communication path to provide input or share concerns outside established battle rhythm events
The space acquisition workforce is experiencing some challenges using the Adaptive Acquisition Framework to tailor acquisition strategies to the unique characteristics of each space acquisition.	Train and familiarize acquisition professionals on tailoring the different acquisition pathways, depending on type of capability, mission, and so forth.

Creating a Shared Vision for Transitioning Capabilities into Operations

As noted in Chapter 4, the space acquisition community lacks evolutionary roadmaps that identify opportunities to insert technologies into operations. In addition, we found a lack of a shared vision regarding criteria for how and when technologies should be fielded into the broader operational architecture. As a result, the outcomes of experiments, innovation, rapid capability development initiatives, and even PoRs are not being efficiently fielded (or are not fielded at all) to support operations. By adopting the architectural concepts discussed in Chapter 4 and repeatable processes for transitioning technology, both the space acquisition and user communities will have a clear, common vision for how and when capabilities should be integrated into the operational architecture. As part of developing a repeatable approach to integrating technology into operations, there is a need for consistent user (or operational) participation with acquisition professionals and developers to develop and review the capability roadmaps. Similarly, there is a need for regular updates from the program offices and materiel leaders, who are accountable for the implementation architectures, to allow capability and operational architects the ability to monitor and evaluate progress of those implementations. This closed-loop, collaborative process may identify opportunities to integrate new technologies or capabilities, lead to a better understanding of where the acquisition community can make trade-offs, and identify what is required to plug in and effectively use the technology.

Ensuring Resource Availability for Transitioning Innovation to Operations

Funding is key to transitioning technology. The operational community requires funding to sustain and integrate the capability into legacy systems. This is especially challenging for PMs, who must anticipate acquisition considerations, identify appropriate acquisition pathways, articulate capability requirements, and identify additional funding streams while the capability is being developed to ensure the timely transition of the capability into the targeted PoR. And this planning extends beyond the transition itself. For example, PMs must identify adequate funding sources to address any new requirements that stem from the transition or eventual integration into a larger system, while also supporting the future operation and sustainment of the capability.[154] Most importantly, funding needs to be preserved for the timely on-ramping or transitioning of emerging technology into ongoing development projects and operations.

There are additional resource challenges to transitioning capabilities into operations. Scheduling for transitions can be a resource constraint, even when funding is available. Transitioning both too early and too late can result in poor outcomes. Transitioning too late delays the capability such that it may be obsolete when delivered or is delivered significantly late to need. Transitioning before the operator is ready could mean that the capability may not be adopted.[155] This may contribute to acquisition organizations' inability to pick up new technology coming out of the research labs or rapid capability development offices at the right time for implementation and fielding.

As discussed in Chapter 6, we recommend identifying funding sources during USSF's Program Objective Memorandum build and protecting those funds throughout the PPBE process. Both the operational and acquisition communities will have to justify and support the need for these funds. Stating that these funds are reserved to transition new or emerging technologies into operations serves multiple purposes: (1) It demonstrates to Congress that USSF is serious about accelerating technology to address an evolving threat environment, (2) it provides transparency, thereby improving congressional relationships, and (3) it delivers operationally relevant capability to operators more rapidly.

Additionally, to make the best use of available funding, it is important to identify and terminate systems and technologies that are either obsolete or not operationally relevant. This would free up funds for higher priorities and reinforce with Congress that USSF is a good steward of the taxpayers' money.

Developing a Formal Process to Provide Feedback to Acquisition Decisionmakers

Interviewees stated that a formal mechanism is needed to provide ongoing feedback from the operational community into the acquisition process, and we found evidence of growing relationships and communication lines between PEOs and deltas, as shown in our social network analysis of the space acquisitions ecosystem (Figure 7.1). In addition to the internal USSF and DoD forums for discussing pain points and the regular interactions we observed between PEOs, SSIO, and the SpOC mission area teams, USSF has developed forums for linking industry teams to operations. For instance, AFWERX and SpaceWERX sponsor an Open Topic initiative, in which companies and researchers propose technology solutions for the warfighter.[156] In 2021, 19

---

[154] Kim et al., 2020.

[155] Michael J. Sullivan, *Weapon Systems: Prototyping Has Benefited Acquisition Programs, but More Can Be Done to Support Innovation Initiatives*, Government Accountability Office, GAO-17-309, June 27, 2017.

[156] AFRL Public Affairs, 2021.

of the nearly two dozen small businesses that pitched their space- and satellite-related technologies at Open Topic received Phase II SBIR contracts to develop prototypes.[157] Evaluators for projects presented at Open Topic come from outside SpaceWERX and AFWERX, allowing the entire DAF operational and acquisition communities to provide feedback and feel empowered to shape SpaceWERX's portfolio.[158]

However, as described in Chapter 5, these interactions appear to be ad hoc or personality dependent rather than formalized or widespread among the acquisition and operational communities. The space acquisition community and operational community need to develop and formalize a process to incorporate warfighter feedback into acquisition decisionmaking. This requires establishing a lead entity or organization to orchestrate the process. In addition, a semiannual or quarterly battle rhythm event for warfighters and developers to discuss ideas, pain points, and requirements may also be useful. Finally, there needs to be a clear communication path to provide input or share concerns outside established battle rhythm.

Figure 7.1. Relationships Between Operations and Acquisition Organizations

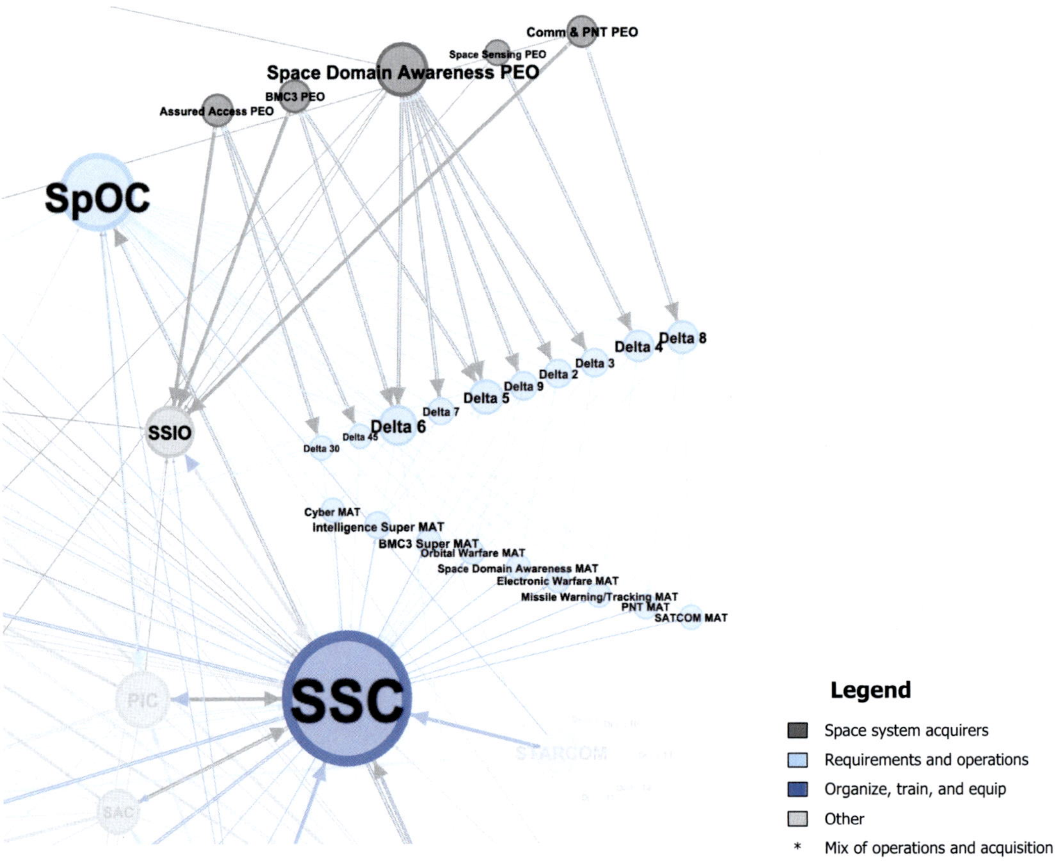

SOURCE: RAND analysis of interviews conducted in spring 2022.

---

[157] AFRL Public Affairs, 2021.

[158] Author interviews with SMEs, January 2022 through July 2022.

NOTE: Node and font size indicates the organization's centrality, calculated as the weighted sum of its interfaces. Line width indicates the strength of the relationship as recorded in our interviews. This diagram depicts organizational relations on the basis of interviews. Therefore, it reflects the as-is state of the ecosystem in spring 2022, prior to the SAE position being filled.

One example of a feedback mechanism is having someone from the program offices sitting with the developer and the warfighter throughout the process (similar to a DevSecOps approach) to ensure that input is incorporated; however, this is a labor-intensive approach that may be difficult to implement because of competing priorities and undermanning, per interviewees. Regardless of the approach, user-specific requirements, such as training and sustainment, are necessary to transition or field capabilities. Consistent warfighter-developer interactions in transitioning capabilities, per lessons learned from software development factories, are necessary to establish feedback loops among user, developer, and acquisition communities.

These formal feedback mechanisms should include ways for the operational community to provide feedback into the requirements development process. As noted in Chapter 4, we heard significant concerns from many in the operational community regarding their ability to influence requirements under the new USSF organizational structure. They expressed concern that, with the standup of SpOC as a field command and the creation of the CSRO, the operational community lost direct operational input into the requirements process. This concern appears to arise from the perception that there is no formal role for SpOC in developing the operational, technical, or implementation requirements for new systems. Interviewees also told us that billets that were previously allocated to the operations community for the purpose of interfacing with the requirement development community were lost during the reorganization. Furthermore, we heard from multiple sources that requirement and analysis capacity and capabilities are insufficient.

Improving United States Space Force Acquisition Workforce's Use of the Adaptive Acquisition Framework

In interviews, the space acquisition workforce discussed challenges with using the newer AAF to tailor acquisition strategies to the unique characteristics of each space acquisition. As discussed in Chapter 1, the AAF is a set of tailored pathways meant to assist the acquisition workforce in planning and executing an acquisition strategy that is more appropriately tailored to the unique characteristics of a program. While use of these pathways has promise, interviewees discussed many challenges associated with transitioning prototypes and rapidly developed capabilities. First and foremost is a lack of training for PEO and PMs in the different acquisition pathways. However, second is a lack of organizational experience and knowledge of best practice in how to transition between pathways. Third is the very real challenge of transitioning a technology that was designed as a short-term gap filler into a long-term strategic capability.

DoDI 5000.02 allows for prototypes or capabilities to be developed using new acquisition pathways, such as the MTA pathway, for rapid fielding.[159] Section 804 of the NDAA for fiscal year 2016 provides DoD the authority to rapidly prototype and field capabilities under MTA authorities.[160] The goal of the MTA pathway is to develop or acquire capabilities that are not yet mature enough to become standalone programs but that may be transitioned (or fielded) within five years from the start date of an MTA program. MTA also authorizes transitioning prototypes and capabilities through a rapid fielding pathway if a set of established criteria are met.[161] Unlike Federal Acquisition Regulation-governed programs, MTA programs are not subject to JCIDS or DoDD 5000.01, *The Defense Acquisition System*, unless specifically provided in the program's implementing guidance.[162] The characteristics of the MTA pathway make it a good option for the space acquisition workforce to use to acquire prototypes of space capabilities, but the workforce needs to understand the challenges associated with transitioning capabilities from prototypes to PoRs to fielding.

MTAs may or may not become an official line item in the defense budget. For example, when the MTA pathway is used to minimally develop a capability before rapid fielding, and that capability simply fills a gap, there may not be a need to transition that capability to a longer-term program.[163] However, successful projects that need to be institutionalized should be transitioned to PoRs. One option is to create a new PoR in the major capability pathway that is governed by the federal Acquisition Regulation and subject to the requirements outlined in DoDD 5000.01 and the JCIDS process.[164] Another option is to transition prototypes or capabilities into a major defense acquisition program or an existing PoR as a component of a larger system that also may reside in the major capability pathway.[165]

When integrating into an existing PoR, the challenge is understanding how individual components integrate into a complex system.[166] If the original experiment was not designed to

---

[159] DoDI 5000.02, 2022.

[160] Public Law 114-92, National Defense Authorization Act for Fiscal Year 2016, Section 804, Middle Tier of Acquisition for Rapid Prototyping and Rapid Fielding, November 25, 2015.

[161] These criteria include: (1) The prototype meets a high-priority warfighter need or reduces the life cycle cost of a weapon system, (2) the original prototyping project was successfully completed, and (3) the prototype was demonstrated in a relevant environment. See Defense Acquisition University, undated-d; Defense Acquisition University, "Middle Tier of Acquisition (MTA) Rapid Fielding," webpage, undated-b.

[162] Ellen M. Lord, "Middle Tier of Acquisition (Rapid Prototyping/Rapid Fielding) Interim Authorities and Guidance," memorandum for secretaries of the military departments, Office of the Under Secretary of Defense for Acquisition and Sustainment, April 16, 2018; DoDD 5000.01, 2022.

[163] Defense Acquisition University, "Middle Tier of Acquisition (MTA)," webpage, undated-a.

[164] Defense Acquisition University, undated-d; DoDD 5000.01, 2022; DoDI 5000.02, 2022. See also Chairman of the Joint Chiefs of Staff Instruction 5123.01, *Charter of the Joint Requirements Oversight Council and Implementation of the Joint Capability Integration and Development System*, Joint Staff, October 30, 2021.

[165] Defense Acquisition University, undated-d.

[166] Shelton et al, 2021.

interoperability standards or to align with the concept of operations used by the larger system, integration may be neither simple nor prudent. Because integration is so critical to transitioning component capabilities into existing PoRs and into legacy capabilities, the ability of the PM to conduct transition and acquisition strategy planning, collaborate with organizational and functional SMEs, and identify requirements prior to transitioning between pathways will have a significant impact on the successful transition of the capability.[167]

Given the relatively recent approval to use nontraditional acquisition pathways within the AAF, such as MTA in 2019 or the software acquisition pathway in 2020, personnel with expertise in these acquisition approaches are lacking. We recommend additional training for space acquisition professionals on tailoring the different acquisition pathways depending on type of capability, mission, and so forth. Understanding how best to employ these pathways and associated methods, particularly such software development methods as DevSecOps, requires specific training that many DoD acquisition professionals do not yet have.[168] Personnel with the skillsets to plan and manage the acquisition of large complex systems (such as BMC2), which requires both an understanding of the possible acquisition pathways and agile software development methods, are especially rare.[169] Skill sets mentioned by our interviewees include a need for acquisition professionals to be more tech savvy to better understand the benefits of different acquisition pathways and to know how to properly budget for the adoption of prototypes, experiments, or rapidly developed capabilities.

---

[167] Defense Acquisition University, "Selecting and Transitioning Pathways," webpage, undated-c.

[168] Jon Ludwigson, *DoD Space Acquisitions: Including Users Early and Often in Software Development Could Benefit Programs*, Government Accountability Office, GAO-19-136, March 18, 2019.

[169] Chaplain, 2019a; see also Bonnie L. Triezenberg, Jason M. Ward, Jonathan Cham, Devon Hill, Sean Robson, and Jeff Fourman, *The Composition and Employment of Software Personnel in the U.S. Department of Defense: An Initial Analysis*, RAND Corporation, RR-A520-1, 2020.

Chapter 8. Conclusions

As the newest armed service, USSF is still settling on its organizational design and how best to align and execute its missions most effectively and efficiently. Over the course of this study, USSF continued to evolve, resulting in several impactful organizational changes.

One change with significant impact on our effort was the nomination and confirmation of a USSF SAE in May 2022. Although this position was mandated by the fiscal year 2020 NDAA, its vacancy affected the roles and responsibilities of existing positions.

A second organizational change with significant impact on our research findings and recommendations was the reorganization of SMC to SSC and the transition of the SSC commander from a PEO to an SoS integrator. These changes require the USSF acquisition community to operate in a new paradigm.

Because this research occurred during this time of significant change in the space acquisition community, our research is focused on how space acquisition functions today and not on what might or might not be dictated in future policies or regulations. This transition necessarily causes ambiguity for USSF and other stakeholders (e.g., industry, Congress, other services, and government agencies). The findings in the previous chapters make note of some of this ambiguity, and the associated recommendations are intended to add clarity in the name of improving space acquisition's integration and synchronization to deliver capabilities more rapidly instead of through disparate systems.

Table 8.1 summarizes our findings and recommendations for five areas of interest corresponding to Chapters 3 through 7 in this report.

Underlying all of these recommendations is a need to develop and promote a USSF culture that is more risk tolerant and enterprise-focused. These recommendations are intended to align the roles and responsibilities of space acquisition from an enterprise perspective, reinforced by the concept of unity of command, and a potential office of primary responsibility and office of collateral responsibility are provided for each. We do not find a natural priority among these issues—each of our chapters deals with a significant issue, and there is no natural ordering among them (i.e., there is no logic where one or more of these issues should be addressed prior to the others). To accommodate USSF resource constraints, in selected cases we have recommended short-term changes that require minimal resources to implement, while the resources for longer-term changes are being acquired and those cases are so noted.

Table 8.1. Summary of Findings and Recommendations

Findings	Recommendations	Office of Primary Responsibility	Office of Collateral Responsibility
Budget Resiliency			
Budget instability at the program level and budget inflexibility overall make it difficult to execute and deliver integrated and synchronized capabilities.	Create resource-loaded roadmaps of mission capability architectures and conduct sensitivity analyses to better understand and prepare for the impacts of budget actions, such as continuing resolutions, changes to funding levels, or program cancellations.	SAF/SQ	SWAC, PEOs, PMs
• Resilience to budget instability has not historically been a consideration when partitioning mission capability between USSF program elements.	• Use the outputs of the analyses to enable better communication of mission capability trade-offs related to funding choices.		
	• To mitigate the impacts of budget instability, incorporate modularity into the budgetary design of program elements and capability roadmaps.	SAF/SQ	PEOs, PMs
• The relative inflexibility of DoD's Planning, Programming, Budgeting, and Execution system makes it difficult to field capabilities to counter rapidly evolving threats.	• To mitigate the impacts of budget inflexibility, create program elements that aggregate planning items to cover a significant portion of a mission capability.	SAF/SQ	FM, PEOs, PMs
• There is insufficient understanding of the economic interactions between USSF and the broader space market and how budget instability impacts the space defense industrial base.	• Develop tools, processes, and methods to assess (1) the resilience of USSF capability roadmaps to changes in the industrial base and (2) the impact of USSF acquisition strategies and/or budgeting decisions on the health of the overall market economy that sustains the base.	SAF/SQ	SWAC, SSIO
More-transparent congressional interaction will be needed if USSF is to mitigate the impact and magnitude of budget instability.	Short term: • Implement a cultural change that values more open communication with Congress and builds alignment across USSF (field commands and acquisition community), Department of the Air Force, the Pentagon, and the greater U.S. space enterprise.	LL	SAF/SQ, SWAC, SpOC, STARCOM
	• Continue to improve USSF workforce education on budget processes involving the larger (DoD, White House, and Congress) financial management community.	SecAF	LL

Findings	Recommendations	Office of Primary Responsibility	Office of Collateral Responsibility
	Long term: • As USSF budget performance improves and matures, request that Congress increase the threshold for reprogramming authority to allow the Secretary of the Air Force to more easily reprogram funds between program elements in a mission capability, with advice from the Chief of Space Operations and the Special Acquisition Executive.	SecAF	CSO, LL, SAF/SQ
Architectural Focus			
There are different aspects of mission, system of systems, and systems architecture with responsibilities spread among multiple organizations.	Define *architecture* and use that definition to document USSF mission statements and charters.	SAF/SQ	SWAC
The space acquisition workforce does not have a consistent understanding of architecture and the roles and responsibilities that are related to architecture.	Educate the USSF workforce on architecture vision, definitions, responsible offices, and relationships among systems and missions within the architecture.	SAF/SQ	STARCOM
There is insufficient system of systems design talent in USSF and the Special Acquisition Executive needs reliable access to that talent. It remains to be seen if splitting the available talent between Space Warfighting Analysis Center, Office of the Assistant Secretary of the Air Force for Space Acquisition and Integration, and Space Systems Integration Office is a sustainable strategy.	Short term: • Clarify Space Warfighting Analysis Center's and Space Systems Integration Office's role in defining system of systems architecture and advising the Service Acquisition Executive and the Office of the Assistant Secretary of the Air Force for Space Acquisition and Integration. While the Office of the Assistant Secretary of the Air Force for Space Acquisition and Integration may choose to outsource its need for system-of-systems engineering talent, advice should not be filtered through an external organization.	SAF/SQ	SWAC, SSIO
	Long term: • Conduct a workforce study to determine how to create and sustain organic USSF system of systems and systems engineering expertise.	SAF/SQ	STARCOM
Alignment of USSF Organizations			
Given the newness of USSF, many acquisition-related organizations lack defined and approved mission statements or charters. This leads to a	Formalize and approve mission statements to clarify and document	SAF/SQ	PEOs, SSC/CC

Findings	Recommendations	Office of Primary Responsibility	Office of Collateral Responsibility
general lack of understanding across the workforce regarding accountabilities, authorities, and touch points.	• the acquisition chain of command versus organize, train, and equip responsibilities for the following organizations and sub-organizations: Office of the Assistant Secretary of the Air Force for Space Acquisition and Integration, Program Executive Officers, Space Systems Integration Office, Space Development Agency, and Space Rapid Capabilities Office		
	• the accountabilities and authorities (if any) for the Space Acquisition Council, Program Integration Council, and any warfighter integration councils	SAF/SQ	SSC/CC
	• the roles of the Office of the Assistant Secretary of the Air Force for Space Acquisition and Integration Mission Area Team versus the Space Operations Command Mission Area Teams	SAF/SQ	SpOC
	• the role of Space Training and Readiness Command in (1) the generation of system requirements to support tactics, techniques, and procedures and (2) the developmental or operational testing of systems prior to fielding.	STARCOM	SpOC, SQ
There is no standard partitioning of USSF missions or mission areas (i.e., portfolios of capability). Joint Publication 3-14, Office of the Assistant Secretary of the Air Force for Space Acquisition and Integration, Space Operations Command, and Space Systems Command all partition the missions differently.	Harmonize the functional constructs to the extent reasonable and standardize mission area terminology.	SAF/SQ	SpOC, SWAC
While we found many venues for communications and feedback pathways between operators and acquirers, they are not clearly and consistently defined.	Continue to improve operator and acquirer communication processes: • Provide a known and consistent forum for operators to provide feedback to senior acquisition leaders (e.g., formalize the pain point process).	SAF/SQ	SpOC
	• At the program level, draft a formal user agreement between the user community and the acquisition program office (akin to what is currently being used in the Adaptive Acquisition Framework software pathway).	PEOs	Delta/CCs, PMs
Harnessing Space Innovation and Experimentation			
Significant experimentation is ongoing in multiple organizations that are devoted to innovation, but this experimentation is not visible centrally or across organizations.	Require organizations involved in space acquisition to improve visibility into their efforts and ensure capability roadmaps include all experiments and innovations.	SAF/SQ	PEOs, PMs

Findings	Recommendations	Office of Primary Responsibility	Office of Collateral Responsibility
Acquirers are incentivized at the individual program level instead of the enterprise level, which limits the willingness to incorporate new or emerging technology.	Implement a cultural change that values enterprise and mission success over personal or individual program success: • Align Program Executive Officer and Program Manager incentives with desired outcomes.	SQ	STARCOM
	• Develop more flexible and mission- or enterprise-focused program requirements—for example, using operator pain points or broad thrust areas.	CSRO	SWAC, PEOs, PMs
Innovations are typically not accompanied by incorporation funding.	• Develop criteria for when to terminate aging systems and delayed programs and require justifications for exceptions, i.e., make it okay to terminate programs • Reallocate funding from terminated or underperforming activities to higher priorities, i.e., make it okay to give up funding.	SAF/SQ	SpOC
There may be insufficient acquisition personnel with the skills and experience to readily adapt innovation and experimentation into programs of record.	Investigate options for recruiting and retaining acquisition talent.	STARCOM	SAF/SQ
Transitioning Technology from the Space Acquisition Community to the Operational Community			
A roadmap identifying opportunities to insert technology into operations does not exist.	In addition to creating the capability roadmaps discussed in Chapter 4, create a clear shared vision between acquisition and user communities as to how and when technologies should be integrated into the broader capability architecture.	SAF/SQ	SpOC
Resources needed to transition capability into operations are underallocated and current incentives can lead to the development of obsolete capabilities or capabilities that are never transitioned.	As discussed in Chapter 6: • Preserve funding and timing to on-ramp or rapidly transition technology that emerges during acquisition into ongoing development projects and/or operations. • Establish a process to stop funding capabilities and technologies that fail to meet operational needs.	SAF/SQ	SpOC

Findings	Recommendations	Office of Primary Responsibility	Office of Collateral Responsibility
As noted in Chapter 5, while we found many venues for communications and feedback pathways between operators and acquirers, they are not clearly and consistently defined.	Develop and formalize a process to incorporate warfighter feedback into acquisition decisionmaking as described in Chapter 5. This includes • establishing a lead entity and/or organization to orchestrate the process • establishing a semi-annual or quarterly battle rhythm event for warfighters and developers to discuss ideas, pain points, and requirements • establishing a clear communication path to provide input or share concerns outside established battle rhythm events.	SAF/SQ	SpOC
The space acquisition workforce is experiencing some challenges using the Adaptive Acquisition Framework to tailor acquisition strategies to the unique characteristics of each space acquisition.	Train and familiarize acquisition professionals on tailoring the different acquisition pathways depending on type of capability, mission, and so forth.	SAF/SQ	STARCOM

NOTES: CSO = Chief of Space Operations; CSRO = Chief Strategy and Resourcing Officer; Delta/CC = Commander of Space Operations Command Delta; FM = Financial Management ; LL = Legislative Liaison; PE = program element; PEO = Program Executive Officer; PM = Program Manager; SAF/AQ = Office of the Assistant Secretary of the Air Force for Space Acquisition and Integration; SecAF = Secretary of the Air Force; SpOC = Space Operations Command; SSC/CC = Commander Space Systems Command; SSIO = Space Systems Integration Office; STARCOM = Space Training and Readiness Command; SWAC = Space Warfighting Analysis Center.

Appendix A. Methodology

This appendix describes the various parts of the methodology used in this analysis. Our methodology included a literature review, discussions with SMEs, and the mapping of the current state of the USSF acquisition ecosystem.

Literature Review

We undertook a broad literature review at the start of the project to gain insight into space acquisition policies, processes, and programs across DoD. Many documents were selected for review, including from prior analyses conducted by FFRDCs, DoD (e.g., Defense Acquisition University and service-specific analyses), and GAO; congressional reporting (e.g., Congressional Research Service and Congressional Budget Office); and trade literature relevant to options for improving integration and synchronization. We also pulled together the latest DAF, USSF, and DoD-relevant policies to inform the roles and responsibilities regarding synchronization and integration in space. A review of the literature was conducted and relevant details were captured and organized. Key points were extracted from sources that directly addressed the synchronization and/or integration of space acquisition policies, processes, and/or programs. There was a focus on points relating to several common problem characteristics and themes pertaining to space acquisition within DoD and across other organizations where relevant. Specifically, management and oversight, authority and responsibility for space acquisition, horizontal and vertical integration, threats to space systems in an evolving environment, use of the AAF, requirements processes, space programs and technologies and their life cycles, and PPBE emerged as potential issue areas that would require further research.

A few key Insights were developed from the literature review that provided areas of focus for finding ways to improve the integration and synchronization of space acquisition and fielding. From the literature, it became evident that synchronization across organizations engaged in space acquisition was needed but lacking, as was the effective integration of space systems from the acquisition community through to the user community. For over a decade, GAO has consistently been calling out the challenges of space acquisition, and RAND reports from recent years discuss this issue as well. Many issues relating to space acquisition are well known and well documented. However, less has been written about such topics as USSF organization and acquisition and the AAF, given that implementation has happened in the past several years. The literature review provided a foundation of the current and changing organization of USSF, use of the AAF, and current and future space systems and programs. This foundation helped inform a set of questions for the various interviewees.

Interviews

We conducted more than 60 semistructured interviews with SMEs and senior decisionmakers throughout DoD, other federal government agencies, and the private sector. Government interview participants represented four main communities: primary space system acquirers; those responsible for requirements and oversight; those responsible for organizing, training, and equipping the space community; and those who play additional supporting roles to the space community.

The space system acquirers came from the SSC, Space RCO, SpaceWERX, and SDA. We chose our interview subjects from these organizations on the basis of their experience (1) as PMs, materiel leaders, and senior decisionmakers at program offices; (2) as functional leads for engineering, contracting, financial management, and test and evaluation; (3) in launch and range operations; (4) as integrators; and (5) as space architects. We also interviewed professionals from the industry and FFRDCs who work with space system acquirers. Interview subjects from these organizations provided insights for our study from their perspectives as decisionmakers and experts in acquisition functional areas such as engineering and contracting.

Requirements interview participants came from USSPACECOM, USSF Deltas, SpOC, and SWAC. Participants from these organizations provided insight into (1) the processes of operational and technical requirements generation for the space area of operations; (2) the processes of capability and resource integration; (3) space operations, doctrine, and wargaming; and (4) the needed capabilities that space systems acquirers fulfill on the basis of the developed force design. Other organizations that contribute to space requirements included terrestrial combatant commands (e.g., USAF Central Command), who are the primary users of traditional space services.

Oversight interview participants came from SAF/AQ and the Office of the Under Secretary of Defense for Acquisition and Sustainment. Participants from these organizations provided insight into (1) acquisition and sustainment policy, including architecture and interoperability; and (2) guidance and governance on the basis of that policy.

Finally, codeveloper interview participants included individuals from MDA, NRO, and NOAA. Participants from these organizations provided insight on policy, processes, and methods related to building and operating (1) systems for U.S. missile defense using space-based sensors as critical inputs, (2) systems that support both DoD and the IC, and (3) U.S. operational environmental satellites, respectively.[170]

---

[170] Our interview plan also included the Defense Information Systems Agency, which builds and operates the systems that provide DoD communications worldwide with satellite communications as the backbone of those systems, and U.S. Navy, U.S. Army, and U.S. Air Force end-user equipment acquisition offices, which build the communications and navigation equipment that access USSF-provided space services. However, we were unable to interview these organizations.

Table A.1 lists the organizations whose members participated in our interviews and briefly describes those organizations' role in USSF and, thus, their area of contribution to our study. All interviews were conducted with the understanding that we would not attribute any comments to anyone we interviewed. We did not provide the interview questions to the interviewees prior to the interviews, but we briefly described the project and the intent of the interviews in our invitations.

Table A.1. Summary of Participating Organizations

Organizational Category	Summary
Space system acquirers	• Space Systems Command: provides systems for enduring missions using a capability evolution approach • Space Rapid Capabilities Office: rapidly integrates existing technology solutions to meet emerging and immediate mission needs • Space Development Agency: runs experiments that leverage commercial or other low-cost space capabilities in new ways • Air Force Research Library: leads discovery, research, and development of affordable warfighting technologies for air, space, and cyber forces • SpaceWERX: focuses on growing the space industrial base by fostering space startups • Space CAMP: platform and application development software factory focused on continuous development and deployment of USSF mission applications to the warfighter • Industry: Inflection Space Strategies, Loverro Consulting, Science Applications International Corporation (SAIC), Northrup Grumman, Maxar • Federally Funded Research and Development Center: Aerospace Corporation
Requirements	• U.S. Space Command: primary requirements generator for the space area of operations • USSF Deltas: focus on executing complex space missions as part of joint operations • Space Warfighting Analysis Center: charged with force design • Space Operations Command: generates, presents, and sustains combat-ready intelligence, cyber, space, and combat support forces and serves as a USSF service component to U.S. Space Command
Oversight	• Office of the Assistant Secretary of the Air Force for Space Acquisition and Integration: leads the synchronization and acquisition of Department of the Air Force space systems and programs • Office of the Under Secretary of Defense for Acquisition and Sustainment: provides guidance and governance related to acquisition and sustainment policy, processes, and methods
Codevelopers	• Missile Defense Agency: builds and operates the systems for U.S. missile defense • National Reconnaissance Office: builds and operates space systems that support both DoD and intelligence community needs • National Oceanic and Atmospheric Administration: builds and operates U.S. operational environmental satellites and provides data and information services

SOURCES: Air Force, "Air Force Research Laboratory," webpage, October 2021; Air Force, "Space Acquisition and Integration," webpage, undated; interviews conducted with USSF officials between February and July 2022; LinkedIn, "Space CAMP," webpage, undated; National Oceanic and Atmospheric Administration, "Satellites," webpage, undated-b; Space Operations Command, homepage, undated; USSF, homepage, undated.

Stakeholder Mapping

The military space acquisition ecosystem has undergone significant change between 2020 and 2022 with the standup of USSF and congressionally mandated changes to space acquisition authorities. To understand the current state of the USSF acquisition ecosystem (i.e., the connections between the various stakeholders in space acquisition), we mapped approximately 90 different stakeholders in USSF acquisition and the relationships between those organizations. We used the content from the stakeholder interviews and recent organizational change announcements to define the organizations. We then graphed the existing relationships, as described in the interviews, using social network analysis tools to visually convey the results. It is important to note that the relationships we graph are those for which we have evidence are currently operational, not those that are called out in policy and doctrine. It is also important to note that this is a representation as of spring 2022 and that these relationships are in flux as USSF matures. See Figure A.1. for the full mapping of space stakeholders.

Figure A.1. Space Acquisition Stakeholder Map

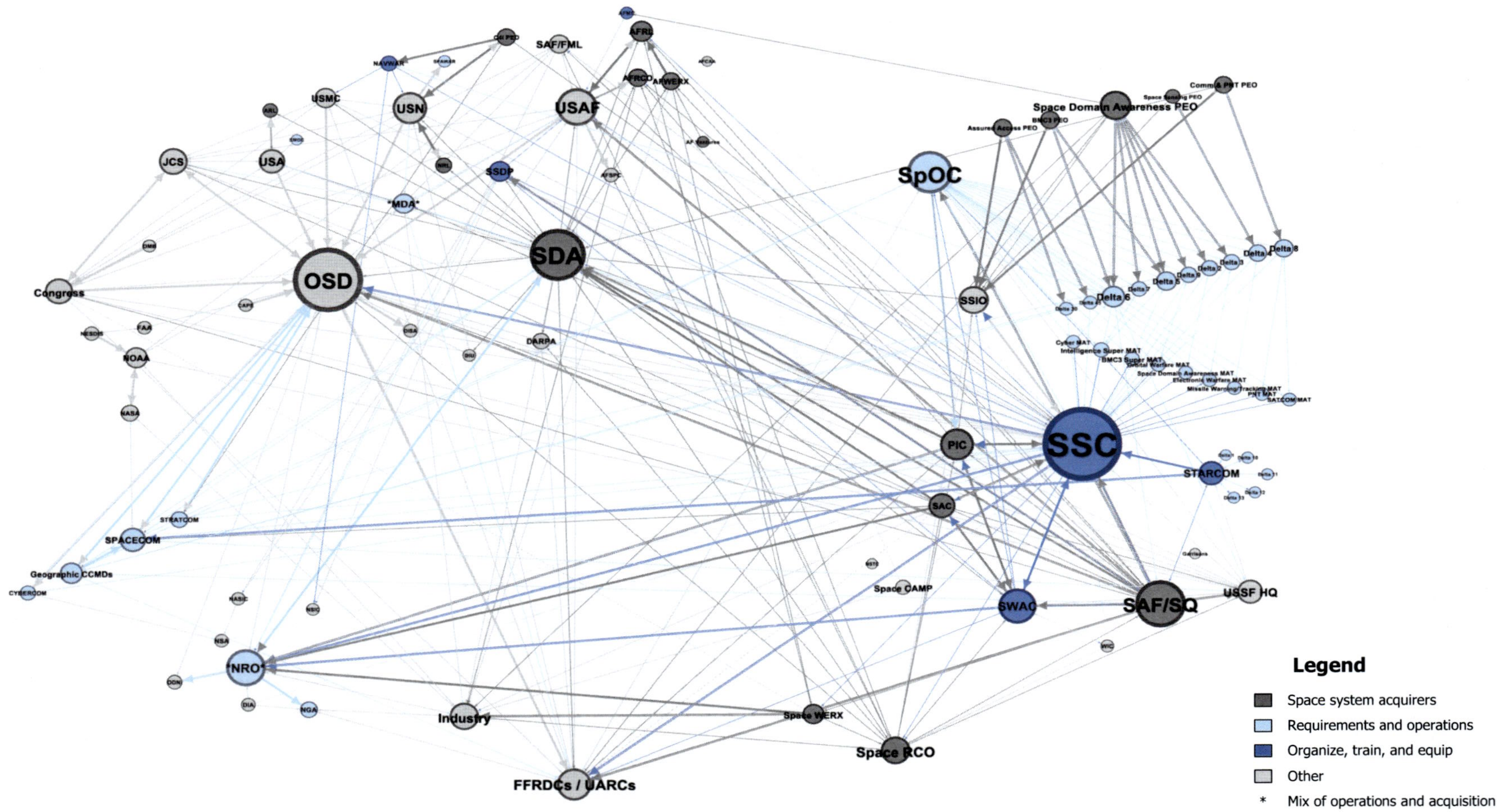

SOURCE: RAND analysis of interviews conducted in spring 2022.
NOTE: Node and font size indicates the organization's centrality, calculated as the weighted sum of its interfaces. Line width indicates the strength of the relationship as recorded in our interviews. This diagram depicts organizational relations on the basis of interviews. Therefore, it reflects the as-is state of the ecosystem in spring 2022, prior to the SAE position being filled.

We collected qualitative data from each interview that pertained to the relationships between organizations involved in space acquisition to understand who was talking to and/or otherwise interacting with whom and the frequency of those interactions. These data were then structured as a graph of nodes and interfaces and visualized using the Gephi social network analysis tool.[171] Figure A.1 consists of approximately 90 nodes connected by 521 edges. Each node represents an organization, department, agency, or office. These nodes were broken down into four groups:

- organizations with formal acquisition authority (dark grey nodes)
- organizations with organize, train, and equip responsibility but no formal acquisition authority (dark blue nodes)
- organizations in the operational community (light blue)
- all other organizations (light grey)

To further assess the relative strength of relationships, we ascribed weighted values to the edge connections between nodes on the basis of information received in the interviews. Those weights are

- thick lines representing strong relationships, i.e., the organizations engage on space acquisition at least weekly
- moderate lines representing relationships that are neither strong nor weak, i.e., the organizations engage multiple times a year
- thin lines representing relationships that are weak, i.e., the organizations that rarely engage
- arrows pointing in the direction of the communication flow in areas where communications seemed primarily unidirectional. In weighting the graph, lines without arrows are counted as two interactions (i.e., one interaction in each direction).

Using these weights allows us to see which stakeholders are most involved in space acquisition (i.e., those nodes with the thickest lines attached to them). Node size is proportional to the number of connections as well as the weights of those connections. Thus, large nodes indicate organizations that frequently engage with other space acquisition stakeholders, have numerous connections with other stakeholders, or both. For example, while NRO has strong relationships with SSC, the PIC, and SDA, it only has 21 edges and thus relatively lower involvement than OSD, which has 47 edges but three strong relationships with FFRDCs, SSC, and SAF/SQ. It is important to note that the width of the lines represent frequency and not quality of interactions. While it is common in social network analysis to term nodes with many strong connections as

---

[171] Gephi is a data analysis tool that allows users to interact with the data and manipulate the structures, shapes, and colors of nodes and connections (i.e., the definition of a graph) to reveal patterns. While Gephi provides layout algorithms to arrange the graph such that the proximity and location of the nodes is indicative of specific mathematical and statistical properties of the graph, we did not use these features and chose to place the nodes on the basis of whether an organization was within USSF (nodes on the right in Figure A.1), DoD, industry, the IC, FFRDCs and university-affiliated research centers, and the federal government. For more information on Gephi and social network analysis, see Martin Grandjean, "GEPHI—Introduction to Network Analysis and Visualization," video, October 14, 2015.

having *influence*, we have largely avoided that word to describe our graphs. This is because *influence* in DoD acquisitions is perhaps more tightly tied to funding than to social interactions and may be more closely related to quality of interactions than to their frequency.

The map shown in Figure A.1 positions organizations on the basis of their affiliations rather than by network metrics, such as closest distance or centrality. Thus, no inferences should be made about influence or quality of interactions on the basis of the proximity of organizations on the map to each other. Instead, we grouped USSF organizations to the right of the map, with other DoD organizations on the top left, other U.S. government organizations on the bottom left, and all others at the bottom center. In future, when USSF organizations and the acquisition ecosystem are more settled and mature, layouts on the basis of network metrics could provide additional insight but, in the current environment, we feel that more sophisticated network analysis techniques are inappropriate.

The stakeholder map underwent multiple iterations involving multiple levels of review with internal project SMEs and the principal investigators, as well as with other space SMEs at RAND to ensure that an accurate depiction of the qualitative data was collected during interviews.

Limitations

The interview protocol (see Appendix B) was not explicitly designed to provide inputs to the space acquisition stakeholder map; put another way, we did not explicitly ask "How frequent are your interactions with the following list of 90 organizations?" Instead, we asked with whom they collaborated and using what mechanisms. This allowed us to organically elicit insights to produce graphs illustrating current and, ideally, *significant* relationships. With more than 90 entities on the graph but just over 60 interviews, there are clearly some areas where our data are incomplete. Relationships may be missing or their strength may be underestimated.[172] Taking a less structured approach to the interviews prioritizes the collection of less explicitly biased social network data over completeness. A more explicit approach in which interviewees directly rate their interactions yields completeness but has been shown to introduce significant bias if

---

[172] It is unlikely that the frequency of interactions is overestimated. Although the interviews ask about collaboration—a word generally viewed as being desirable—we did not specifically ask about frequency or indicate that frequent interactions were desirable. Our choice of the word collaboration may, however, have led to underestimation. For instance, in Chapter 3 we report that interactions with Congress are relatively infrequent and appear to be mediated through OSD. This could be because few of our interviewees regard congressional and USSF interactions as having a collaborative partnership.

Similarly, relationships with industry—those who actually implement and deliver space systems—appear weaker than might be expected. In reviewing our notes regarding interactions with industry, we find that no one pointed to industry as being a contributor to or having solutions for the lack of integration and synchronization of space systems. Thus, the size of the industry node is likely not due to its lack of influence on acquisitions per se but due to its lack of influence on the issues of integration and synchronization of acquisitions.

organizations believe there are incentives related to having their organization perceived as "well connected" or influential.[173]

Another limitation of our study is our reliance on largely self-reported data. As an example of bias in self-reporting, we noted that some organizations reported healthy relationships with Congress, while others engaged in in-depth discussions of problems related to their relationships with Congress. Because congressional relations are a significant topic in our Chapter 3 discussion on budgetary resilience, we undertook a small effort to obtain additional evidence as to the quality of the relations. For USSF acquisitions–related testimony before the HASC, SASC, and the Senate and House Appropriations Committees for Defense from 2020 to late 2022, we obtained audio and written transcripts. Our observation was that the tone of the testimony was highly correlated to committee: the HASC was perhaps the most intense and extensive. Therefore, for testimony before the HASC, we assigned three researchers to read through and/or listen to the testimony to note adversarial exchanges and rate them on a 3-point intensity scale. The researchers performed this task independently for the same testimony and then compared notes to align their interpretation of adversarial intensity. The remaining transcripts were then split up for analysis and we reviewed the overall statistics. We found that organizations who had reported healthy relationships had significantly more adversarial exchanges with HASC members than those who had reported difficulties. In fact, we found some exchanges with those who reported difficulty as being very positive and cordial, with no adversarial exchanges. This can, perhaps, be explained if those organizations who reported difficulties with Congress spent additional time and resources to ensure that those interactions went smoothly. For those who reported healthy relations that were objectively judged as adversarial, it may be that the interviewees consider adversarial exchanges to be healthy. In either case, the disparities between self-reported interactions and the analysis of the actual interactions emphasize the limitations of relying only on self-reported data.

---

[173] Other researchers have found that if there is an implied "goodness" in being well connected, respondents to explicit questions will overemphasize their connections. For instance, Tobias Stark reports an instance of students reporting their entire class as being their "best friend" (Tobias H. Stark, "Collecting Social Network Data," in David L. Vannette and Jon A. Krosnick, eds., *The Palgrave Handbook of Survey Research*, Springer, 2018). Even in work settings, it has been noted that making questions too explicit "might tempt some respondents to name everyone on their organization's roster, even those with whom they communicate only rarely, which renders their responses useless" (Filip Agneessens and Giuseppe (Joe) Labianca, "Collecting Survey-Based Social Network Information in Work Organizations," *Social Networks*, Vol. 68, January 2022, p. 42).

Appendix B. Interview Protocols

As described, we conducted the interviews for this project on a not-for attribution basis, a method that RAND values as an often critical element of objective research. Prior to conducting interviews, we determined that interviewees would likely have perspectives relevant to one of three focus areas of the research on the basis of their professional backgrounds and current roles. To enable the team to make the best use of interviewees' time, roles, and levels of expertise, we developed three distinct interview protocols. Each protocol contained a section of unique core questions to focus on collecting information most relevant to each viewpoint: big picture, acquisition pathways, and stakeholder and user community. Introductory questions about the interviewees' backgrounds and concluding questions of a more general nature were kept consistent across the interview protocols, while a distinct section of questions within each protocol allowed for the comparison of perspectives and responses to those particular questions that are relevant to each focus area.

In addition, we felt that the perspective of PEOs was unique, given their leadership roles and authorities, and therefore crafted a fourth interview protocol for PEOs to drive questioning toward insights that included perspectives that were crosscutting of the research areas in our focus. What follows are the interview protocols for the three main focus areas of our research, as well as the fourth protocol used for interviews of PEOs. These protocols are presented without revision.

Common Elements for All Protocols

Introduction and Informed Consent

My name is <insert name> and I'm part of a team in Project Air Force at the RAND Corporation researching the various approaches used in space acquisition. Our sponsor, the Office of the Assistant Secretary of the Air Force for Space Acquisition and Integration (SAF/SQ), wants to improve how the United States Space Force (USSF) delivers timely, end-to-end space capabilities to the joint force. Our goal is to understand how acquisition can support the full range of space capabilities; from single-purpose systems (e.g., weather sensing; launch services) to multipurpose systems of systems capabilities composed of multiple satellite constellations, payloads, ground stations, user equipment, operations and data processing centers (e.g., SIBRS, protected tactical SATCOM). Given your expertise, we thank you for sharing your insights.

This interview will last about an hour. Your participation is completely voluntary, and the information you provide will be treated confidentially. You are free to stop the interview at any time or skip a question you prefer not to answer. We will only be asking for your opinions in a professional capacity and will not ask for personal information beyond your professional role.

The risks of participation are minimal. We will not attribute comments to individuals without prior permission. We are following procedures to ensure that there will not be any inadvertent release of information including storing all data in a secure, password-protected location, not using individuals' names in any notes, and making sure no one is identifiable by inference.

If you have any questions about this interview, please feel free to contact the principal investigators. Do you have any questions before we proceed?

Background Questions

Please tell us about your current position and experience within the space community.

- What positions have you held?
- What organization do you currently work for within the space community?
- What is your current role and responsibilities in acquiring space capabilities?

Concluding Questions

1. What keeps you up at night?
    a. IF NEEDED: In other words, what is the single most important takeaway we should leave this conversation with?
2. Are there areas that you feel we have missed and should consider in our work?
3. Is there anyone else who you think we should talk to?

Interview Protocol for Big Picture Task

Big Picture Questions

1. What do you see are the critical challenges in developing and acquiring space capabilities (and we mean end-to-end capability vice systems)?
    a. What about **integration**—do you see that as a challenge? Why/why not?
    b. What about **synchronization**—do you see that as a challenge? Why/why not?
    c. Are problems external to the space community?
    d. How much do you manage internally and how much requires collaboration outside your organization?
2. What practices are you using currently to anticipate or alleviate these challenges?
3. Which programs that you oversee have tailored acquisition strategies? What are the strategies and how are they tailored?
4. What is working about acquisition approach and where are there challenges?
    a. What changes would you make if you were able to make changes?
    b. Were any of these characteristics unique (or uniquely important) because they are space systems?

Interview Protocol Focused on Acquisition Pathways

Acquisition Pathway Decision and Coordination

1. **What is the acquisition strategy for your current acquisition program and what approaches/tools are you using to manage its execution?**
    a. What was your acquisition approach and why was it selected?
    b. Can you share documents that describe your acquisition approach?
    c. What is working about your acquisition approach and where are there challenges?
        i. Were you able to tailor your acquisition approach?
        ii. What changes would you make if you were able to make changes?
        iii. Were any of these characteristics unique (or uniquely important) because this is a space system?
2. **(For the functional organizations) How did the selection of an acquisition strategy impact your execution?**
    a. What input did you have in the acquisition strategy decision?
    b. If you had your preference, what approach would you have recommended and why?
3. **How and with whom do you collaborate outside of your organization? (e.g., with other USSF organizations, other services, other federal government, industry, or foreign partners, etc.)**
    a. What issues/information do you integrate and synchronize between organizations?
    b. What means do you use to integrate with other organizations (formal or informal)?
    c. Are there examples from your experience of integration across organizations being particularly successful? What about particularly challenging? And why?

Interview Protocol Focused on Stakeholder and User Community

Stakeholder and User Community Perspective Questions

1. Thinking about your current role, do you currently interact with the acquisition community?
    a. IF YES: In what capacity and frequency?
    b. IF NO: Why not, did you in the past, do you think you will in the future, etc.?
2. How is the acquisition process collaborative and/or responsive to emerging needs?
3. How might the acquisition process evolve or change to better meet stakeholder needs?
4. Do you use a process, other than JCIDS, to set requirements for rapid or agile acquisition? If so, what is that process?
    a. How do you prioritize requirements?
    b. How do you decide which ones you are willing to trade for cost and/or schedule?
    c. How do you balance the trade-offs between cost, schedule, and performance?
5. How is your participation in rapid or agile acquisition different than in traditional acquisition?
6. How does the user community contribute to synchronizing and integrating space capabilities?

Interview Protocol Focused on PEO Roles and Responsibilities

PEO-Related Questions

1. **What is it about space that makes it different than acquisition for other domains?**
2. **What are the key challenges you are facing in acquiring space capabilities? And what are some recommendations that you have for alleviating these challenges?**
3. The current Air Force Instruction (AFI) describing the accountabilities of PEOs centers on cost, schedule and performance; often referred to as the iron triangle. The AFI does not reference architecture, integration, resilience, or interoperability. **How do you balance the need to build architectural resiliency and interoperability within that cost/schedule/performance triangle?**
4. **Beyond the formal JCIDS and the Warfighter Integration councils, how would you better integrate acquisitions with…**
    a. the operational deltas?
    b. TTP development?
    c. test and evaluation?
    d. sustainment and maintenance personnel?
    e. the terrestrial CoComs or other DoD Services and Agencies who will employ/use space systems?
5. **Do you feel you own the accountability for the architecture/roadmap for your mission area? How do you see your accountabilities vs.**
    a. SWAC
    b. The System of Systems Integration Office
    c. The Allied integration office
    d. The Commercial integration office
    e. The SpOC Mission Area Team
    f. The SpOC deltas
    g. The STARCOM deltas
6. **How are decisions made about which agency to use to procure various pieces of the overall architecture for your mission area (SDA, Space RCO, SSC)?**
7. **How do you decide which acquisition pathways to use for the various parts of the architecture?**
    a. What are your thoughts regarding transitioning pieces of the architecture between acquisition pathways?
8. **What are the organizations you interact with most both inside and outside of the acquisition community?**
9. **What was the budget process used this last year? Who did you interact with to ensure your mission area received its fair share? How do you see the process changing in the coming years? What will the PEO role be?**

Abbreviations

AAF	Adaptive Acquisition Framework
AFRL	Air Force Research Laboratory
ATLAS	Advanced Tracking and Launch Analysis System
BMC2	Battle Management Command and Control
C2	command and control
CSO	Chief of Space Operations
CSRO	Chief Strategy and Resourcing Officer
DAF	Department of the Air Force
DMSP	Defense Meteorological Satellite Program
DoD	Department of Defense
DoDD	Department of Defense Directive
FFRDC	federally funded research and development center
FORGE	Future Operationally Resilient Ground Evolution
GAO	Government Accountability Office
GPS	Global Positioning System
HASC	House Armed Services Committees
IC	intelligence community
ISR	intelligence, reconnaissance, and surveillance
JADC2	Joint All-Domain Command and Control
JCIDS	Joint Capabilities Integration and Development System
LOE	line of effort
MAT	Mission Area Teams
MDA	Missile Defense Agency
MTA	middle tier of acquisition
NC3	nuclear command, control, and communications
NDAA	National Defense Authorization Act
NOAA	National Oceanic and Atmospheric Administration
NRO	National Reconnaissance Office
OSD	Office of the Secretary of Defense
OT&E	organize, train, and equip
PAE	Portfolio Acquisition Executive
PAF	Project AIR FORCE
PE	program element
PEO	Program Executive Officer
PIC	Program Integration Council

PM	Program Manager
PNT	position, navigation, and timing
PoR	program of record
PPBE	Planning, Programming, Budgeting, and Execution
SAC	Space Acquisition Council
SAE	Service Acquisition Executive
SAF/SQ	Office of the Assistant Secretary of the Air Force for Space Acquisition and Integration
SASC	Senate Armed Services Committees
SATCOM	satellite communications
SBIR	Small Business Innovation Research
SDA	Space Development Agency
SMART	specific, measurable, attainable, realistic, and timely
SMC	Space and Missile Systems Center
SME	subject-matter expert
SoS	system of systems
Space RCO	Space Rapid Capabilities Office
SpOC	Space Operations Command
SSC	Space Systems Command
SSIO	Space Systems Integration Office
STARCOM	Space Training and Readiness Command
SWAC	Space Warfighting Analysis Center
UDL	Unified Data Library
USAF	U.S. Air Force
USD/A&S	Under Secretary of Defense for Acquisition and Sustainment
USSF	United States Space Force
USSPACECOM	U.S. Space Command

References

AFRL—*See* Air Force Research Laboratory.

Agneessens, Filip, and Giuseppe (Joe) Labianca, "Collecting Survey-Based Social Network Information in Work Organizations," *Social Networks*, Vol. 68, January 2022.

Ahlbin, Jon, "MDA University Outreach QS Radiation Hardness Assurance for Parts and Materials (QSR) Division," briefing slides, Missile Defense Agency, 2021.

Air Force, "Space Acquisition and Integration" webpage, undated. As of July 11, 2022: https://www.safsq.hq.af.mil

Air Force, "Air Force Research Laboratory," webpage, October 2021. As of July 31, 2022: https://www.af.mil/About-Us/Fact-Sheets/Display/Article/104463/air-force-research-laboratory/

Air Force Instruction 63-101/20-101, *Integrated Life Cycle Management*, U.S. Department of the Air Force, June 30, 2020, incorporating Change 1, November 23, 2021.

Air Force Research Laboratory Public Affairs, "SpaceWERX Ready to Propel Space Innovation," Air Force Research Laboratory, August 25, 2021.

Albon, Courtney, "US Space Force Considers Purchasing Weather Data as a Service," *DefenseNews*, January, 11, 2022a.

Albon, Courtney, "US Space Force Rapid Capabilities Office to Deliver First Project This Year," *C4ISRNET*, July 7, 2022b.

"Altamira Awarded an $8 Million+ Space Mission Data Processing Award by USSF," *SatNews*, December 10, 2020.

"Analysing Business Process Pain Points to Get Better Requirements," Aotea Studios, 2011.

Air Force Chief Software Officer, "Space CAMP," webpage, December 19, 2019. As of July 9, 2022:
https://software.af.mil/softwarefactory/spacecamp/

Birkler, John, Mark V. Arena, Irv Blickstein, Jeffrey A. Drezner, Susan M. Gates, Meilinda Huang, Robert Murphy, Charles Nemfakos, and Susan K. Woodward, *From Marginal Adjustments to Meaningful Change: Rethinking Weapon System Acquisition*, RAND Corporation, MG-1020-OSD, 2010. As of June 22, 2022:
https://www.rand.org/pubs/monographs/MG1020.html

Carberry, Sean, "Congress Alarmed over Innovation Funding," *National Defense Magazine,* July 6, 2022.

Chairman of the Joint Chiefs of Staff Instruction 5123.01, *Charter of the Joint Requirements Oversight Council and Implementation of the Joint Capability Integration and Development System*, Joint Staff, October 30, 2021.

Chaplain, Cristina T., *Space Acquisitions: DoD Delivering New Generations of Satellites, but Space System Acquisition Challenges Remain*, testimony before the Subcommittee on Strategic Forces, Committee on Armed Services, U.S. Senate, Government Accountability Office, GAO-11-590T, May 11, 2011.

Chaplain, Cristina T., *Defense Space Acquisitions: Too Early to Determine If Recent Changes Will Resolve Persistent Fragmentation in Management and Oversight*, Government Accountability Office, GAO-16-592R, July 27, 2016.

Chaplain, Cristina T., *Space Acquisitions: DoD Continues to Face Challenges of Delayed Delivery of Critical Space Capabilities and Fragmented Leadership*, testimony before the Subcommittee on Strategic Forces, Committee on Armed Services, U.S. Senate, Government Accountability Office, GAO-17-619T, May 17, 2017.

Chaplain, Christina T., *Space Acquisitions: DoD Faces Significant Challenges as It Seeks to Address Threats and Accelerate Space Programs*, Government Accountability Office, GAO-19-482T, March 27, 2019a.

Chaplain, Cristina T., *Space Command and Control: Comprehensive Planning and Oversight Could Help DoD Acquire Critical Capabilities and Address Challenges*, Government Accountability Office, GAO-20-146, October 30, 2019b.

Chief Information Officer, Department of Defense, "DoDAF Viewpoints and Models," webpage, undated-a. As of May 8, 2023: https://dodcio.defense.gov/Library/DoD-Architecture-Framework/dodaf20_viewpoints/

Chief Information Officer, Department of Defense, "DoDAF Viewpoints and Models: Capability Viewpoint," webpage, undated-b. As of May 8, 2023: https://dodcio.defense.gov/Library/DoD-Architecture-Framework/dodaf20_capability/

Chief Information Officer, Department of Defense, "DoDAF Viewpoints and Models: Operational Viewpoint," webpage, undated-c. As of May 8, 2023: https://dodcio.defense.gov/Library/DoD-Architecture-Framework/dodaf20_operational/

Chief Information Officer, Department of Defense, "DoDAF Viewpoints and Models: Services Viewpoint," webpage, undated-d. As of May 8, 2023: https://dodcio.defense.gov/Library/DoD-Architecture-Framework/dodaf20_services/

Collins, K., D. Kazdan, and N. A. Frissell, "Ham Radio Forms a Planet-Sized Space Weather Sensor Network," *Eos*, February 9, 2021.

Conway, Melvin E., "How Do Committees Invent?" *Datamation*, Vol. 14, No. 5, April 1968.

DAF—*See* Department of the Air Force.

Daniels, Seamus P., and Kaitlyn Johnson, "What the Continuing Resolution Means for Defense and Space Force Funding in FY 2021," Center for Strategic and International Studies, October 9, 2020.

Defense Acquisition University, "Middle Tier of Acquisition MTA," webpage, undated-a. As of May 8, 2023:
https://aaf.dau.edu/aaf/mta/

Defense Acquisition University, "Middle Tier of Acquisition (MTA) Rapid Fielding," webpage, undated-b. As of May 25, 2023:
https://aaf.dau.edu/aaf/mta/fielding/

Defense Acquisition University, "Selecting and Transitioning Pathways," webpage, undated-c. As of May 25, 2023:
https://aaf.dau.edu/aaf/selecting-a-pathway/

Defense Acquisition University, "Transition Program," webpage, undated-d. As of May 8, 2023:
https://aaf.dau.edu/aaf/mta/prototyping/transition/

Demarest, Heidi Brockmann, *US Defense Budget Outcomes: Volatility and Predictability in Army Weapons Funding*, Palgrave Macmillan, 2017.

Department of Defense, *Defense Space Strategy Summary*, June 2020.

Department of Defense, *Developmental Test and Evaluation*: *FY 2012 Annual Report*, March 2013.

Department of Defense, *DoD Dictionary of Military and Associated Terms*, November 2021.

Department of Defense, "Space Development Agency Makes Award for 126 Satellites to Build Tranche 1 Transport Layer," press release, February 28, 2022a.

Department of Defense, *Summary of the Joint All-Domain Command and Control (JADC2) Strategy*, March 2022b.

Department of Defense Directive 5000.01, *The Defense Acquisition System*, Office of the Under Secretary of Defense for Acquisition and Sustainment, September 9, 2020, incorporating Change 1, July 28, 2022.

Department of Defense Directive 5205.07, *Special Access Program (SAP) Policy*, U.S. Department of Defense, July 1, 2010, incorporating Change 2, February 4, 2020.

Department of Defense Instruction 5000.02, *Operation of the Adaptive Acquisition Framework*, Office of the Under Secretary of Defense for Acquisition and Sustainment, January 23, 2020, incorporating Change 1, June 8, 2022.

Department of the Air Force, *Department of Defense Fiscal Year (FY) 2022 Budget Estimates: Air Force Justification Book Volume 1 of 1, Research Development, Test and Evaluation, Space Force*, May 2021.

Department of the Air Force, *United States Space Force Vision for Satellite Communications (SATCOM)*, January 23, 2020.

Dimarogonas, James, Jasmin Léveillé, Jan Osburg, Shane Tierney, Bonnie L. Triezenberg, Graham Andrews, Bryce Downing, Muharrem Mane, and Monica Rico, *Universal Command and Control Language Early System Engineering: Performance Effects of a Universal Command and Control Standard*, RAND Corporation, RR-A744-2, 2023. As of May 23, 2023:
https://www.rand.org/pubs/research_reports/RRA744-2.html

DoD—*See* Department of Defense.

DoDD—*See* Department of Defense Directive.

DoDI—*See* Department of Defense Instruction.

Doran, George T., "There's a S.M.A.R.T. Way to Write Management's Goals and Objectives," *Management Review,* Vol. 70, No. 11, November 1981.

Drezner, Jeffrey A., *Competition and Innovation Under Complexity*, RAND Corporation, RP-1386, 2009. As of August 22, 2022:
https://www.rand.org/pubs/reprints/RP1386.html

Drezner, Jeffrey A., and Michael Simpson, *Exploring Parallel Development in the Context of Agile Acquisition: Analytical Support to the Air Superiority 2030 Enterprise Capability Collaboration Team*, RAND Corporation, RR-1808-AF, 2017. As of June 22, 2022:
https://www.rand.org/pubs/research_reports/RR1808.html

Edwards, Jane, "Lt. Gen. Michael Guetlein on System of Systems Integration Within Space Systems Command," *ExecutiveGov*, March 7, 2022.

Erwin, Sandra, "Space Force Procurement Command Trying to Bring 'Unity of Effort' to Space Programs," *SpaceNews*, February 23, 2022a.

Erwin, Sandra, "Space Force to Reorganize Its Acquisition Command to 'Focus on the Threat,'" *SpaceNews*, March 4, 2022b.

Erwin, Sandra, "U.S. Space Force to Step Up Protection of Satellite Ground Systems in the Wake of Russia's Cyber Attacks," *SpaceNews*, May 19, 2022c.

Erwin, Sandra, "Space Force Delivers First of Two U.S. Payloads to Launch on Space Norway's Arctic Broadband Mission," *SpaceNews*, June 9, 2022d.

Erwin, Sandra, "New Head of Space Force Acquisitions Looks to Get Back to Basics," *SpaceNews*, June 24, 2022e.

Erwin, Sandra, "Senate NDAA Directs Independent Review How the Space Force Acquires Technologies," *SpaceNews*, July 21, 2022f.

Erwin, Sandra, and Brian Berger, "A Race Against Time to Replace Aging Military Weather Satellites," *SpaceNews*, May 24, 2021.

Field, Elizabeth A., *Defense Budget: DoD Has Adopted Practices to Manage Within the Constraints of Continuing Resolutions*, Government Accountability Office, GAO 21-541, September 13, 2021.

Gephi, homepage, undated. As of July 11, 2023:
https://gephi.org/

Goya, Blythe, "18th Space Control Squadron Recognized for Innovation," Vandenberg Space Force Base, Combined Force Space Component Command Public Affairs, November 29, 2021.

Grandjean, Martin, "GEPHI—Introduction to Network Analysis and Visualization," video, October 14, 2015. As of July 11, 2023:
https://www.youtube.com/watch?v=GXtbL8avpik

Greenwalt, William, and Dan Patt, *Competing in Time: Ensuring Capability Advantage and Mission Success Through Adaptable Resource Allocation*, Hudson Institute, February 2021.

Hadley, Greg, "Study: Combine Missile Warning, Tracking Constellations into One Multi-Orbit System," *Air and Space Forces Magazine*, June 7, 2022.

Hitchens, Theresa, "New Space BMC2 System to Debut at Corona: Raymond," *Breaking Defense*, September 30, 2019.

Hitchens, Theresa, "Newest Sats Launched by DoD Include Jammer-Evading, Classified Payloads," *Breaking Defense*, July 6, 2022a.

Hitchens, Theresa, "Companies Knock, but Space Force Acquisition Portal 'Front Door' Remains Closed," *Breaking Defense*, July 7, 2022b.

Hitchens, Theresa, "Rocket Science: How Space Force Acquisition Works, with Many Players and Dual Hats," *Breaking Defense*, July 27, 2022c.

Hitchens, Theresa, "NRO, Air Force May Co-Fund Future Space-Based ISR Tech: Kendall," *Breaking Defense*, August 1, 2022d.

Joint Doctrine Note 2-19, *Strategy*, Joint Chiefs of Staff, December 10, 2019.

Joint Publication 3-14, *Space Operations*, Office of the Joint Chiefs, April 10, 2018, incorporating Change 1, October 26, 2020.

Kim, Yool, Guy Weichenberg, Frank Camm, Brian Dougherty, Thomas C. Whitmore, Nicholas Martin, and Badreddine Ahtchi, *Improving Acquisition to Support the Space Enterprise Vision*, RAND Corporation, RR-2626-AF, 2020. As of June 22, 2022: https://www.rand.org/pubs/research_reports/RR2626.html

Koster, Kim, "What Your Schedule Isn't Telling You," paper presented at the PMI Global Congress 2011, Dallas, Tex., 2011.

Landreth, James M., "Through DoD's Valley of Death," Defense Acquisition University, February 1, 2022.

LinkedIn, "Space CAMP," webpage, undated. As of July 31, 2022: https://www.linkedin.com/company/spaceforce-spacecamp

Lofgren, Eric, "Data: How DoD Reprogramming Has Changed (1961–2019)," Acquisition Talk, May 10, 2022.

Lord, Ellen M., "Middle Tier of Acquisition (Rapid Prototyping/Rapid Fielding) Interim Authorities and Guidance," memorandum for secretaries of the military departments, Office of the Under Secretary of Defense for Acquisition and Sustainment, April 16, 2018.

Lorell, Mark A., Julia F. Lowell, and Obaid Younossi, *Evolutionary Acquisition: Implementation Challenges for Defense Space Programs*, RAND Corporation, MG-431-AF, 2006. As of June 22, 2022:
https://www.rand.org/pubs/monographs/MG431.html

Ludwigson, Jon, *Space Acquisitions: Challenges Facing DoD as It Changes Approaches to Space Acquisitions*, testimony before the Subcommittee on Strategic Forces, Committee on Armed Services, House of Representatives, Government Accountability Office, GAO-16-471T, March 9, 2016.

Ludwigson, Jon, *DoD Space Acquisitions: Including Users Early and Often in Software Development Could Benefit Programs*, Government Accountability Office, GAO-19-136, March 18, 2019.

Ludwigson, Jon, *Space Acquisitions: DoD Faces Challenges and Opportunities with Acquiring Space Systems in a Changing Environment*, testimony before the Subcommittee on Strategic Forces, Committee on Armed Services, House of Representatives, Government Accountability Office, GAO-21-520T, May 24, 2021a.

Ludwigson, Jon, *Missile Warning Satellites: Comprehensive Cost and Schedule Information Would Enhance Congressional Oversight*, Government Accountability Office, GAO-21-105249, September 22, 2021b.

Ludwigson, Jon, *GPS Modernization: Better Information and Detailed Test Plans Needed for Timely Fielding of Military User Equipment*, Government Accountability Office, GAO-22-105086, May 9, 2022.

Mayer, Lauren A., Mark V. Arena, Frank Camm, Jonathan P. Wong, Gabriel Lesnick, Sarah Lovell, Edward Fernandez, Phillip Carter, and Gordon T. Lee, *Prototyping Using Other Transactions: Case Studies for the Acquisition Community*, RAND Corporation, RR-4417-AF, 2020. As of November 1, 2022:
https://www.rand.org/pubs/research_reports/RR4417.html

McCormick, William, "DoD Plans to Replace DMSP Weather Satellites Within Five Years," *ExecutiveGov*, May 24, 2021.

McGarry, Brendan W., "Defense Primer: DoD Transfer and Reprogramming Authorities," Congressional Research Service, IF11243, version 6, December 21, 2021.

McGarry, Brendan W., and Heidi M. Peters, "Defense Primer: Future Years Defense Program (FYDP)," Congressional Research Service, IF10831, version 10, December 15, 2021.

Miller, Amanda, "Experimental IR Satellite Heads to GEO to Advance Hypersonic Missile Warning and Track," *Air and Space Forces Magazine*, June 29, 2022.

Mintzberg, Henry, *The Structuring of Organizations*, Prentice Hall, 1979.

"Narrowband vs. Wideband," *JEM Engineering* blog, September 2021. As of July 11, 2023:
https://jemengineering.com/blog-narrowband-vs-wideband/

National Oceanic and Atmospheric Administration, "Defense Meteorological Satellite Program (DMSP)," Office of Satellite and Product Operations, webpage, undated-a. As of July 2022:
https://www.ospo.noaa.gov/Operations/DMSP/index.html

National Oceanic and Atmospheric Administration, "Satellites," webpage, undated-b. As of July 11, 2022:
https://www.noaa.gov/satellites

National Park Service, "Modularity in Resilient Systems," webpage, last updated February 24, 2017. As of July 11, 2023:
https://www.nps.gov/subjects/culturallandscapes/resilientsystems_modularity.htm

Newton, Richard, "Bridging the Valley of Death and Thriving Beyond—Improving Transition Success with Metrics," Defense Acquisition University, July 30, 2022.

NOAA—*See* National Oceanic and Atmospheric Administration.

Orkand Corporation, *Monopsony: A Fundamental Problem in Government Procurement,* Aerospace Research Center, 1973.

Oxford English Dictionary, "experimentation, *n.*," webpage, undated. As of July 17, 2023: https://www-oed-com.dclibrary.idm.oclc.org/view/Entry/66540

Oxford English Dictionary, "experiment, *n.*," webpage, undated. As of July 17, 2023: https://www-oed-com.dclibrary.idm.oclc.org/view/Entry/66530

Oxford English Dictionary, "innovation, *n.*," webpage, undated. As of July 17, 2023: https://www-oed-com.dclibrary.idm.oclc.org/view/Entry/96311

Platform One, homepage, undated. As of August 10, 2022: https://p1.dso.mil

Pope, Charles, "Kendall Details 'Seven Operational Imperatives' and How They Forge the Future Force," Secretary of the Air Force Public Affairs, March 3, 2022.

Project Thor Team, *Outpacing the Threat with an Agile Defense Space Enterprise*, Aerospace Corporation, September 2019.

Public Law 114-92, National Defense Authorization Act for Fiscal Year 2016; Section 804, Middle Tier of Acquisition for Rapid Prototyping and Rapid Fielding, November 25, 2015.

Public Law 116-92, National Defense Authorization Act for Fiscal Year 2020, December 20, 2019.

Reuters, "Clinton Acts to Make GPS More Accurate," *New York Times*, May 2, 2000.

Sawyer, John D., *Missile Defense: Better Oversight and Coordination Needed for Counter-Hypersonic Development*, Government Accountability Office, GAO-22-105075, June 16, 2022.

SDA—*See* Space Development Agency.

Section 809 Panel, "Recommendation 36: Transition from a Program-Centric Execution Model to a Portfolio Execution Model," *Report of the Advisory Panel on Streamlining and Codifying Acquisition Regulations*, Vol. 3, January 2019.

Shelton, William, Cynthia R. Cook, Charles Barton, Frank Camm, Kelly Elizabeth Eusebi, Diana Gehlhaus, Moon Kim, Yool Kim, Megan McKernan, Sydne Newberry, and Colby P. Steiner, *A Clean Sheet Approach to Space Acquisition in Light of the New Space Force*, RAND Corporation, RR-A541-1, 2021. As of June 22, 2022: https://www.rand.org/pubs/research_reports/RRA541-1.html

Simon, Larry, "EM&C and FMI—Taking Small Steps to an Integrated Satellite Architecture," Government Satellite Report, September 10, 2020.

Snyder, Don, James D. Powers, Elizabeth Bodine-Baron, Bernard Fox, Lauren Kendrick, and Michael H. Powell, *Improving the Cybersecurity of U.S. Air Force Military Systems Throughout Their Life Cycles*, RAND Corporation, RR-1007-AF, 2015. As of June 22, 2022: https://www.rand.org/pubs/research_reports/RR1007.html

Space Development Agency, "Battle Management Command, Control, and Communication (BMC3)," webpage, undated. As of July 2022: https://www.sda.mil/battle-management/

Space Development Agency, *Delivering Capabilities*, April 23, 2021.

Space Force Public Affairs, "Japan's Office of National Space Policy Signs Historic MOU with the U.S. Space Force," Space Force News, December 18, 2020.

SpOC—*See* Space Operations Command.

Space Operations Command, homepage, undated. As of July 31, 2022: https://www.spoc.spaceforce.mil

Space Operations Command, *Space Operations Command Mission Brief*, April 21, 2022.

Space Rapid Capabilities Office, "Space Rapid Capabilities Office (Space RCO) Overview," briefing slides, October 14, 2020.

Space Systems Command, "SSC Realignment," United States Space Force, briefing slides, version 14.0, December 8, 2021.

Space Warfighting Analysis Center, *SWAC—Shaping the Force Design of U.S. Space Force*, December 17, 2021.

SpaceWERX, "About Us," webpage, undated. As of August 10, 2022: https://spacewerx.us/about-us/

Stark, Tobias H., "Collecting Social Network Data," in David L. Vannette and Jon A. Krosnick, eds., *The Palgrave Handbook of Survey Research*, Springer, 2018.

Sullivan, Michael J., *Weapon Systems: Prototyping Has Benefited Acquisition Programs, but More Can Be Done to Support Innovation Initiatives*, Government Accountability Office, GAO-17-309, June 27, 2017.

SWAC—*See* Space Warfighting Analysis Center.

Tirpak, John A., "Bunch Says Quick-Reaction Funds Would Help New Tech Cross the 'Valley of Death,'" *Air and Space Forces Magazine*, June 10, 2022.

Triezenberg, Bonnie L., Colby P. Steiner, Grant Johnson, Jonathan Cham, Éder Sousa, Moon Kim, and Mary Kate Adgie, *Assessing the Impact of U.S. Air Force National Security Space Launch Acquisition Decisions: An Independent Analysis of the Global Heavy Lift Launch Market*, RAND Corporation, RR-4251-AF, 2020. As of July 19, 2022: https://www.rand.org/pubs/research_reports/RR4251.html

Triezenberg, Bonnie L., Jason M. Ward, Jonathan Cham, Devon Hill, Sean Robson, and Jeff Fourman, *The Composition and Employment of Software Personnel in the U.S. Department of Defense: An Initial Analysis*, RAND Corporation, RR-A520-1, 2020. As of August 2, 2022: https://www.rand.org/pubs/research_reports/RRA520-1.html

U.S. Code, Title 10, Section 2273a, Space Rapid Capabilities Office.

U.S. Code, Title 10, Section 3103, Civilian Management of the Defense Acquisition System.

U.S. Code, Title 10, Section 9016, Assistant Secretaries of the Air Force.

U.S. Code, Title 10, Section 9021, Space Acquisition Council.

U.S. Code, Title 10, Section 9086, Space Development Agency.

United States Space Force, homepage, undated. As of July 11, 2022: https://www.spaceforce.mil

United States Space Force, "Space Based Space Surveillance," fact sheet, March 22, 2017.

USSF—*See* United States Space Force.

Werner, Debra, "Air Force Taps Applied Technology Associates and Teledyne Brown to Produce Space Weather Sensors," *SpaceNews*, January 29, 2020.

Williams, Shara, Jeffrey A. Drezner, Megan McKernan, Douglas Shontz, and Jerry M. Sollinger, *Rapid Acquisition of Army Command and Control Systems*, RAND Corporation, RR-274-A, 2014. As of November 1, 2022: https://www.rand.org/pubs/research_reports/RR274.html

Wolfe, Frank, "Timeline for Fielding Unified Data Library Uncertain, as U.S. Space Force Continues Prototyping," *DefenseDaily*, April 14, 2022.

Wolfe, Frank, "U.S. Space Force Unfunded Priorities List Includes $431 Million for Programs to Provide 'Warfighting Punch,'" *Defense Daily*, June 7, 2021.

Wong, Jonathan P., Obaid Younossi, Christine Kistler LaCoste, Philip S. Anton, Alan J. Vick, Guy Weichenberg, and Thomas C. Whitmore, *Improving Defense Acquisition: Insights from Three Decades of RAND Research*, RAND Corporation, RR-A1670-1, 2022. As of July 7, 2022: https://www.rand.org/pubs/research_reports/RRA1670-1.html

Yonekura, Emmi, Brian Dolan, Moon Kim, Krista Romita Grocholski, Raza Khan, and Yool Kim, *Commercial Space Capabilities and Market Overview: The Relationship Between Commercial Space Developments and the U.S. Department of Defense*, RAND Corporation, RR-A578-2, 2022. As of July 13, 2022: https://www.rand.org/pubs/research_reports/RRA578-2.html